Robots, Androids, and Animatrons

Robots, Androids, and Animatrons

12 Incredible Projects You Can Build

John Iovine

McGraw-Hill

New York   San Francisco   Washington, D.C.   Auckland   Bogotá
Caracas   Lisbon   London   Madrid   Mexico City   Milan
Montreal   New Delhi   San Juan   Singapore
Sydney   Tokyo   Toronto

...in-Publication Data

...ns : 12 incredible projects you can

Includes index.
ISBN 0-07-032804-8
    1. Robotics—Amateurs' manuals.   I. Title.
TJ211.5.I58   1997
629.8'92—dc21

97-34715
CIP

McGraw-Hill

A Division of The McGraw·Hill Companies

8 9 0   FGR/FGR   0 2 1 0

ISBN 0-07-032804-8

The sponsoring editor for this book was Scott Grillo, the editing supervisor was Sally Glover, and the production supervisor was Sherri Souffrance. It was set in ITC Century Light per the EL3 Design by Paul Scozzari and Kim Sheran of McGraw-Hill's Professional Book Group composition unit, Hightstown, N.J.

Printed and bound by Quebecor/Fairfield.

McGraw-Hill books are available at special quantity discounts to use as premiums and sales promotions, or for use in corporate training programs. For more information, please write to the Director of Special Sales, McGraw-Hill, 11 West 19th Street, New York, NY 10011. Or contact your local bookstore.

 This book is printed on recycled, acid-free paper containing a minimum of 50 percent recycled, de-inked fiber.

Dedication
To Ellen, my wife;
James, my son; and
AnnaRose, my daughter.
With love.

Acknowledgments

I would like to thank some of the people who helped make this book possible; Matt Wagner; my agent at Waterside Productions; Scott Grillo, who tried to keep me on schedule; and Sally Glover for a great job of editing.

Contents

Index

Introduction

There are many interesting and fun things to do in electronics, and one of the most enjoyable is building robots. Not only do you employ electronic circuits and systems, but they must be merged with other technologies. Building a robot from scratch involves the following:

- ☐ Power supply systems
- ☐ Motors and gears for drive and motion control
- ☐ Sensors
- ☐ Artificial intelligence

Each one of these technologies has numerous books dedicated to its study. Naturally, a comprehensive look at each technology isn't possible in one book, but we will touch upon these areas, and you will gain hands-on knowledge and a springboard for future experimentation.

Robotics is an evolving technology. There are many approaches to building robots, and no one can be sure which method or technology will be used 100 years from now. Like biological systems, robotics is evolving following the Darwinian model of survival of the fittest.

You're not alone when you become a robotist. I was surprised to learn that there are many people, government organizations, private organizations, competitions, and clubs devoted to the subject of amateur robotics. NASA has the most advanced robotics systems program I ever saw. Much of the information is free for the asking. If you have Internet access, jump to one of the search engines (Yahoo, Excite, etc.) and search under robotics. You will find the Web sites of many companies, individuals, universities, clubs, and newsgroups dedicated to robotics.

About the Author

John Iovine is the author of several popular TAB titles that explore the frontiers of scientific research. He has written *Homemade Holograms: The Complete Guide to Inexpensive, Do-It-Yourself Holography; Kirlian Photography: A Hands-On Guide; Fantastic Electronics: Build Your Own Negative-Ion Generator and Other Projects;* and *A Step into Virtual Reality.*

In the beginning

THE ORIGIN OF ROBOTICS CAN BE TRACED BACK TO THE ancient Greeks, who created movable statues. Around 270 BC, Ctesibus (a Greek engineer) made organs and water clocks with movable figures.

In the 1770s, Pierre Jacquet-Droz, a Swiss clock maker and inventor of the wrist watch, created three ingenious mechanical dolls. He made the dolls so that each one could perform a specific function: one could write, another could play music on an organ, and the third could draw a simple picture. As sophisticated as they were, the dolls performed all their respective feats using gears, cogs, pegs, and springs. Their purpose was to amuse royalty.

More recently, in 1898, Nikola Tesla built a radio-controlled submersible boat. This was no small feat in 1898. The submersible was demonstrated in Madison Square Garden. Although Nikola Tesla had plans to make the boat autonomus, lack of funding prevented further research.

The word robot was first used in a 1921 play titled *R.U.R.— Rossum's Universal Robots* by Czechoslovakian writer Karel Capek. The word *robot* is a Czech word for workers. The play described mechanical servants, the "robots." When the robots were endowed with emotion, they turned on their masters and destroyed them.

Historically, we have sought to endow human abilities and attributes into inanimate objects that resemble the human form. From this is derived the word *anthrobots*, creating robots in human form.

In times since, robots have become a staple in many science fiction stories and movies. As robots evolved, so did the terminology needed to describe the different robotic forms. So in addition to the old "tin-man" robot, we also have *cyborgs*, which are part human and part machine, and *androids* which are specially built robots designed to be humanlike.

Many people had their first look at a real robot during the 1939 World's Fair. Westinghouse Electric built a robot they called Elektro the Moto Man. Although Electro had motors and gears to move his mouth, arms, and hands, it could not perform any useful work. He was joined on stage by a mechanical dog named Sparko.

Why build robots?

Robots are indispensable in some manufacturing industries. The reason is because the cost per hour to operate a robot is a fraction of what human labor would cost to perform the same function. More than this, once programmed, robots can produce a highly repeatable accuracy that surpasses the most experienced human operator. However, human operators are far more versatile. Humans can switch job tasks easily. Robots are job specific. You wouldn't be able to program a welding robot to start counting parts in a bin.

Today's most advanced robots are really contemporary robotic dinosaurs. Robots are in their infancy stage of evolution. As robots evolve that will become more versatile, emulating the human capacity and ability to switch job tasks easily.

While the personal computer has made an indelible mark in society, the personal robot hasn't made an appearance. Obviously there's more to a personal robot than a personal computer. Robots require a combination of elements to be effective: sophistication of intelligence, movement, mobility, navigation, and purpose.

Purpose

In the beginning, personal robots will focus on a single function (job task) or purpose. For instance, today there are small, mobile robots that can autonomously maintain a lawn by cutting the grass. These robots are solar powered and don't require any training. Underground wires are placed around the lawn perimeter. The robots sense the wire, remain within the defined perimeter, and don't wander off.

Building a useful personal robot is difficult. It's beyond the scope of this book, or for that matter, of any other contemporary book on robotics. So you may ask, "What's the purpose of this book?" Well, in reading this book and building a few robots, you will become part of the ongoing robotics evolution.

Creativity and innovation do not require a degree. Robot building is not restricted to Ph.Ds., college universities, professors, and industrial giants. By playing and experimenting with robots, you can learn many aspects of robotics: artificial intelligence, neural networks, usefulness and purpose, sensors, navigation, articulated limbs, etc. The potential here is to make a contribution to the body of knowledge existing on robotics. And to this end, amateur robotics is making tremendous strides forward, in some cases using a clever design to surpass mainstream robotics development.

As the saying goes, look before you leap. When designing a robot, the first question to ask is, "What is the purpose of this robot?" What will it do and how will it accomplish its task? My dream is to build a small robot that will change my cat's litter box when it needs changing.

This book provides tools: sensors, drive systems, neural nets, and microcontrollers that you can use to build your own robot. Before we begin to build, let's look at some current applications and how robots may be used in the future. The most sophisticated robots today are built by NASA and the U.S. military. NASA's main interest in robotics involves space exploration and telepresence. The military, on the other hand, uses the technology in warfare.

Exploration

NASA routinely sends unmanned robotic explorers where it is impossible to send human explorers. Why send robots instead of humans? Economics. It's much cheaper to send an expendable robot than a human. Humans require an enormous support system to function: breathable atmosphere, food, heat, living quarters, and, quite frankly, most want to live through the experience and return to Earth in their lifetime.

Explorer spacecraft travel through the solar system, where their electronic eyes transmit back to Earth fascinating pictures of the planets and their moons. The Viking probes sent to Mars looked for life and sent back pictures of the Martian landscape. NASA is developing planetary rovers, space probes, spider-legged walking explorers, and underwater rovers. NASA has the most advanced telerobotic program in the world, operating under the Office of Space Access and Technology (OSAT).

NASA estimates that by the year 2004, 50% of extra vehicle activity (EVA) will be conducted through the use of telerobotics.

For a complete explanation of telerobotics and telepresence, see Chapter 9.

Space probes launched from Earth have given us spectacular views of our neighboring planets in the solar system. And in this era of tightening budgets, robotic explorers provide the best value for the taxpayer dollar. Robotic explorer systems can be built and implemented for a fraction of the cost of manned flights. Let's examine a typical case. The Mars Pathfinder Project represents a new generation of small, low-cost spacecraft and explorers.

Mars Pathfinder (Sojourner)

The Mars Pathfinder consists of a lander and rover. It launched from Earth in December of 1996 on board a McDonnell Douglas Delta II rocket to begin its journey to Mars. It arrived on Mars on July 4, 1997.

The Pathfinder did not go into orbit around Mars. Instead, it flew directly into the Mars atmosphere at 17,000 mph (27,000 km/hr, 7.6 km/s). To prevent burning up in the atmosphere, Pathfinder used a combination of a heat shield, parachute, rockets, and airbags. Although the landing was cushioned with airbags, Pathfinder experienced a deceleration of 40 gravities (Gs).

Pathfinder landed in an area known as Ares Vallis. The site is at the mouth of an ancient outflow channel where potentially a large variety of rocks are within reach of the rover. The rocks settled there, being washed down from the highlands at a time when there were floods on Mars. The Pathfinder craft opened up after landing on Mars (see Fig. 1.1) and released the robotic rover.

The rover on Pathfinder is called Sojourner (see Fig. 1.2). Sojourner is a new class of small robotic explorers that are sometimes called microrovers. Sojourner is small, weighing in at 22 lbs (10.5 kg); height is 280 millimeters (mm) (10.9 inches), length is 630 mm (24.5"), and width is 480 mm (18.7"). The rover has a unique six-wheel (Rocker-Bogie) drive system developed by Jet Propulsion Laboratories (JPL) in the late 1980s.

The main power for Sojourner is a solar panel made up of over 200 solar cells. Power output from the solar array is about 16 watts. Sojourner has been exploring the surface of Mars since its landing in July, in 1997.

Previously, this robot was known as Rocky IV. The development of this microrover robot went through several stages and prototypes, including Rocky I, Rocky II, Rocky III, and, of course, Rocky IV.

4

■ **1.1** *Mars Pathfinder. Photo courtesy of NASA*

Both the Pathfinder lander and rover have stereo imaging systems. The rover carries an alpha proton X-ray spectrometer that it uses to determine the composition of rocks. The lander will make atmospheric and meteorological observations and be the radio relay station for the rover.

Mission objectives:

The Sojourner rover itself is an experiment. Performance data from Sojourner will determine how effective microrover explorers are and can be. In addition, the following experiments will be performed.

Experiments

☐ Long-range and short-range surface imaging

☑ Soil mechanics

☐ Mars dead-reckoning sensor performance

☐ Sinkage in Martian soil

■ **1.2** *Sojourner Rover. Photo courtesy of NASA*

☐ Logging vehicle performance data
☐ Rover thermal characterization
☐ Rover imaging sensor performance
☐ UHF link effectiveness
☐ Material abrasion
☐ Material adherence
☐ Alpha proton X-ray spectrometer
☐ Imaging
☐ APXS deployment mechanism
☐ Lander Imaging
☐ Damage assessment

Sojourner is controlled via telepresence by an Earth-based operator. The operator navigates the rover using images obtained from the rover and lander. Because the time delay between the Earth-operator actions and the rover's response will be between 6 and 41 minutes depending on the relative positions of Earth and Mars, Sojourner has onboard intelligence to help prevent accidents, like driving off a cliff.

NASA is continuing development of microrobotic rovers. Small robotic land rovers with intelligence added for onboard navigation, obstacle avoidance, and decision making are planned for future Mars exploration. This robotic system provides the best value per taxpayer dollar.

The latest microrover currently being planned for the next Mars expedition will again check for life. On August 7, 1996 NASA released a statement saying that it believes to have found fossilized microscopic life on Mars. This information has renewed interest in searching for life on Mars, but it is too close to the Pathfinder launch to modify the Sojourner rover to search for life.

Industrial robots—going to work

Robots are indispensable in many manufacturing industries. For instance, robot welders are commonly used in automobile manufacturing. Other robots are equipped with spray painters and paint components. The semiconductor industry uses robots to solder (spot weld) microwires to semiconductor chips. Other robots (called pick and place) insert integrated circuits (ICs) onto printed circuit boards, a process known as *stuffing the board*.

These particular robots perform the same repetitive and precise movements day in and day out. This type of work can become very boring to a human operator. Following operator boredom is fatigue and, with operator fatigue, errors. Production errors reduce productivity, which leads directly to higher manufacturing costs. Higher manufacturing costs are passed along to the consumer as higher retail prices. In a competitive market, the company that provides high-quality products at the best (lower) price succeeds.

Robots are ideally suited for performing repetitive tasks. Robots are faster and cheaper than human laborers. This is one reason that manufactured goods are available at low cost. Robots improve the profit and competitiveness of manufacturing companies. Without robots, many companies would no longer be able to compete in their industries.

Design and prototyping

Some robots are useful for more than repetitive work. Manufacturing companies commonly use computer aid design (CAD), computer aided manufacturing (CAM), and computer numerical control (CNC) machines to produce designs, manufacture components, and assemble machines. These technologies allow an engineer to design a component using CAD and manufacture the design of the board using computer-controlled equipment quickly. Computers assist in the entire process, from design to production.

Hazardous duty

Without risking human life or limb, robots can replace humans in some hazardous duty service (see Fig. 1.3). For example, bomb disposal robots are used in many bomb squads across the nation. Typically, these robots resemble small armored tanks and are guided remotely by personnel using video cameras attached to the front of the robot. Robotic arms can grab a suspected bomb and place it in a bomb-proof box for detonation and/or disposal.

In the future, similar robots can help in toxic waste cleanup. Robots can work in all types of polluted environments, chemical and nuclear—pollution so hazardous that an unprotected human would be killed quickly. It was the nuclear industry that first developed and used robotic arms. Robotic-arms technology was required because it allowed scientists to be located in safe rooms, operating controls for the robotic arms located in hot rooms. The robotic arms handled lethal radioactive materials.

Maintenance

Maintenance robots specially designed to travel through pipes, sewers, air conditioning ducts, and other systems can assist in as-

■ **1.3** *Hazbot. Photo courtesy of NASA*

sessment and repair. A video camera mounted on the robot can transmit video pictures back to an inspecting technician. Where there is damage, the technician can use the robot to facilitate small repairs quickly and efficiently.

Fire-fighting robots

Better than a home fire extinguisher, how about a home fire-fighting robot? This robot will detect a fire anywhere in the house, travel to the location, and put out the fire.

Fire-fighting robots are so attractive that this type of robot has an annual national competition open to all robotists, who try to develop the best robot. The Fire-Fighting Home Robot Contest is sponsored by Trinity College, the Connecticut Robotics Society, and a number of corporations. Typically, the robot becomes active in response to the tone from a home fire alarm. Its job is to navigate through a mock house and locate and extinguish the fire.

Medical robots

Medical robots fall into four general categories. The first category relates to diagnostic testing. In the spring of 1992, Neuromedical Systems Inc. of Suffern, New York, released a product called Papnet. Papnet is a neural network tool that helps cytologists detect cervical cancer quickly and more accurately.

Laboratory analysis of pap smears is a manual task. A technician examines each smear under a microscope, looking for a few abnormal cells among a larger population of normal cells. The abnormal cells are an indicator of patients with a cancerous or precancerous condition. Many abnormal cells are missed due to human fatigue and habituation.

Scientists have been trying to automate this checking process for 20 years using computers with standard rule-based programming. This was not a successful approach. The difficulty is that the classic algorithms could not differentiate between the complex visual patterns of normal cells and abnormal cells.

Papnet uses an advanced image recognition system and neural network. The network selects 128 of the most abnormal cells found on a pap smear for later review by a cytologist.

The Papnet system is highly successful. It recognizes abnormal cells in 97% of the cases. Since the reviewing technician is only looking at 128 cells instead of 200,000–500,000 cells on pap smear, the fatigue factor is greatly reduced. In addition, the time required to review a smear is only one-fifth to one-tenth what it was before. The accuracy improves to a rate of 3% false negatives, as compared to 30–50% for manual searches.

The second medical category relates to telepresence surgery. Using a specially developed medical robot, a surgeon is able to operate on a patient remotely. The robot has unique force-feedback sensors that relate to the surgeon the feel of the tissue underneath the robot's instruments. This technology makes it possible for specialists to extend their talent to remote provinces of the world.

The third category relates to VR and enhanced manipulation. With enhanced manipulation, the surgeon operates on a patient through a robot. The robot translates all of the surgeon's movements. This makes it possible for the surgeon to perform micro-surgical procedures that can only be dreamed about today. For instance, let's suppose the surgeon moves his or her hand 1 inch. The computer would translate that to 1/10" or 1/100" travel. It is now possible for the surgeon to perform delicate, microscopic surgery that would normally be impossible.

Nanotechnology

Nanotechnolgy controls and manipulates matter at the atomic and molecular level. It is the ability to create electronic and mechanical components using individual atoms. These tiny (nano) components can be assembled to make machines and equipment the size of bacteria. IBM has already created transistors, wires, gears, and levers out of atoms.

How does one go about manipulating atoms? Two physicists, Gerd Binnig and Heinrich Rohrer, invented the scanning tunneling microscope (STM). The tip of the STM is very sharp and its positioning exact. In 1990, IBM researchers used an SMT to move 35 xenonatoms on a nickel crystal to spell the company's name, "IBM." The picture of "IBM" written in atoms made worldwide news and was shown in many magazines and newspapers. This marked the beginning of atomic manipulation. IBM continues to improve its nanotechnology, creating electronic and mechanical components, and nanotechrobotics will find many uses in manufacturing, exploration, and medicine.

Nanotech medical bots

Nanotechnology is used to create small, microscopic robots—image robots so small that they can be injected into a patient's bloodstream. The robots travel to the heart and begin removing the fatty deposits, restoring circulation. Or the robots travel to a tumor, where they selectively destroy all cancerous cells. What are now inoperable conditions may one day be cured through nanotechnology.

Another hope of nanomedical robots is to stop and reverse the aging process in humans. Tiny virus-sized nano-bots enter each cell, resetting the cell clock back to 1. Interesting possibilities.

Keep in mind that nanotechnology is a whole new robotics field itself. Macroscopic and microscopic robots will do everything from cleaning your house to building and processing materials. Everyone expects that nanotechnology will be creating new, high-quality materials and fabrics at low cost.

War robots

One of the first applications of robots is war. If forced into a war, robots can help us win, and win fast. Robots are becoming increasingly more important in modern warfare. Drone aircraft can track enemy movements and keep the enemy under surveillance.

The Israeli military used an unmanned drone in an interesting way. The drone was created to create a large radar target. It was flown into enemy airspace. The enemy switched on its targeting radar, allowing the Israelis to get a fix on the radar position. The radar installation was destroyed, making it safe for fighter jets to follow through.

Smart bombs and cruise missiles are other examples of "smart" weaponry. As much as I appreciate Asimov's 3 Laws of Robotics, which principally state that robots should never intentionally harm human beings, war bots are here to stay.

Robot Wars

This leads to an interesting civilian competition held once a year called Robot Wars. Competitors build remote-controlled robots that are matched by weight in one-on-one battle. Winners advance through elimination.

The arena for the competition is 30 by 54 feet of smooth asphalt with 8-foot high walls to protect spectators. For more information on Robot Wars, check the following Web site:
http://www.robotwars.com.

You can reach them by snail mail at:

Robot Wars
LLC
PO Box 936
Fairfax CA 94978

Civilian

Robotic drones and lighter-than-air aircraft (blimps) developed by the military could be put to civilian use monitoring high-crime neighborhoods and traffic conditions. Because the aircraft do not have any human occupants, they can be made much smaller. I feel that robotic blimps will be used more often in comparison to robotic aircraft because blimps will be safer to operate. Aircraft need to move in order to maintain lift, and an out-of-control aircraft drone can become a lethal weapon if it flies into anything. Blimps, on the other hand, are safer because they travel slower and they float gracefully through the air. If surveillance aircraft become reliable enough, they could also be put to use monitoring traffic, warehouses, apartment buildings, and street activity in high-crime areas.

Domestic

Domestic robots could clean windows, do minor home repairs, vacuum the house, clean the upholstery, wash clothes, and change the kitty litter. Should we consider our dishwashers, washing machines, and clothes dryers robots or machines? I think that when they gather the materials that need to be cleaned by themselves, they will have passed the machine stage and entered into robotics.

Robots will find many more uses and niches that aren't thought of today. These applications will not become apparent until robots are so prevalent in society that the application becomes a mixture of availability, imagination, and need.

More uses

Robotic applications and research develop faster than anyone can follow. The Internet is an excellent tool for finding information. The following is a list of creative robotic applications being developed that are posted on the Internet.

Multifunction automated crawling system (MACS)

MACS are being developed to inspect the exterior of aircraft. There are areas of aircraft that are difficult for people to reach. MACS use ultrasonic motors, suction cups for gripping the aircraft surface, computers, and video feeds. More information on MACS can be found on the Internet at:

`http://robotics.jpl.nasa.gov/tasks/macs/homepage.html`

Catching object in flight

A robotic arm has been built that can catch a model airplane in flight. More information can be found at:

`http://www.ai.mit.edu/projects/handarm-haptics/airplane.html`

Ants

Ants are microbots developed at MIT. The microbots are approximately the size of a cubic inch. Despite their small size, they are packed with 19 sensors that include four IR sensors, four light sensors, two bump sensors, five food sensors, one tilt sensor, two mandible position sensors, and one battery voltage sensor.

The IR sensors are used for communication among the ants. These mobile robots can be programmed for various games that involve social behaviors, such as "tag" and "follow the leader." More info on this interesting bots can be found at:

`http://www.ai.mit.edu/projects/ants`

Farming

Industrial farm equipment is being retrofitted with robotic controls. Self-propelled agricultural equipment like combines and tractors have been successfully retrofitted, and more information can be found at:

`http://rec.ri.cmu.edu/projects/DEMETER/dem_95.html`

DRIP

DRIP stands for "dinky robot in pool." These are low-cost underwater robots, and more information can be found at:

`http://www.kipr.org/robots/drip.html`

Ariel

Ariel is an autonomous underwater legged robot that resembles a crab. It is designed to transverse the surf zone of beaches to find and destroy underwater mines. Ariel is sponsored by DARPA (Defense Advanced Research Projects Association). More information can be found at:

`http://www.isr.com/Ariel.html`

Nanomanipulator

This is an interesting robotic system that integrates telepresence to a scanning tunneling microscope (SMT) and hand-operated

13

force-feedback manipulator. The SMT is a microscope and a nanomanipulator. By connecting the SMT to the force-feedback manipulator, an operator can, in real time, perform nano-operations like cleaving virus particles, writing their name using atoms, or building molecule-sized gears. More information can be found at:

`http://www.cs.unc.edu/Research/nano/nanopage.html`

COG

Cog is a humanoid robot being developed at the MIT Artificial Intelligence Lab under the leadership of robotist Rodney Brooks, who also developed the extremely useful subsumption architecture used in many robot insects. The idea in designing a humanoid robot is to encourage humanlike interaction between the robot and its users. Another facet of the technology is learning how to replicate things like human vision, hearing, touch, etc. To this end, MIT has already made advances in understanding binocular vision. More information can be found at:

`http:www.ai.mit.edu/project/cog/Text/cog-robot.html`

Artificial life and artificial intelligence

2

THE EVOLUTION OF ROBOTICS LEADS TO TWO INTERESTING areas, the creation of artifical intelligence and artifical life.

Artifical intelligence

People dream of creating a machine or artifical intelligence (AI) that rivals or surpasses human intelligence. I feel that neural networks are the leading technology for developing and generating artificial intelligence in computer systems. This is in contrast to other computerists, who see expert systems and task-specific rule-based systems (programs) as potentially more viable.

It is undeniable that rule-based computer operating systems (Windows 95, OS/2, DOS, etc.) and rule-based software are valuable and do most of the computer labor today. However, the pattern-matching and learning capabilities of neural networks are the most promising approach to realizing the AI dream.

Recently it had been forecast that large-scale parallel processors using a combination of neural networks and fuzzy logic could simulate the human brain within 10 years. While this forecast may be optimistic, progress is being made toward achieving that goal. Second-generation neural chips are on the market. Recently, two companies (Intel, Santa Clara, California, and Nestor Inc., Providence, Rhode Island) through joint effort created a new neural chip called the Ni1000. Released in 1993, the Ni1000 chip contains 1,024 artificial neurons. This integrated circuit has three million transistors and performs 20 billion integer operations per second.

Evolution of consciousness in artificial intelligence

Consciousness is a manifestation of the brain's internal processes. The evolution of consciousness coincides with the evolution and development of neural structures (the brain) in biological systems. A billion years ago, the highest form of life on Earth was a worm. Let's consider the ancestral worm for a moment. Does its rudimentary (neural structure) intelligence create a form of rudimentary consciousness? If so, then it's akin to an intelligence and consciousness that can be created by artificial neural networks capable of running in today's supercomputers (see Fig. 2.1).

In reality, while the processing power of supercomputers approaches that of a worm, this has not yet been accomplished. The reason is that it is difficult to program a neural network in the supercomputer that can use all of the computer's processing power.

A worm is unquestionably alive, but is it self-aware? Is it simply a cohesive jumble of neurons replaying an ancestral record imprinted within its primordial neural structure, making it no more than a functional biological automaton?

This raises a few questions: Is intelligence conscious? Is consciousness life? Perhaps intelligence has to reach a certain level or

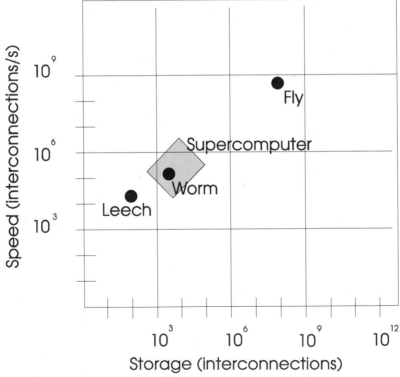

■ **2.1** *Graph showing supercomputer capabilities*

critical mass before consciousness is achieved. In any case, artificial neural networks can and will develop consciousness. Whether the time frame is 10 years or a 1,000 years, both estimates are less than a blink of an eye in an evolutionary time line.

Artificial life

There are a few different areas of artificial life. In general, there are three main areas of artificial life: self-replicating robots, nanorobotics (sometimes self-replicating), and programs (software).

The most evolved type of artificial life on Earth today is programs. No one has created a self-replicating robot, and nano bots are still years away. Therefore, let's limit our discussion to programs for the time being. In AI programs, life exists only as electrical impulses that make up the running program inside a computer. Computer scientists have created artificial life programs that mimic many fuctions of life (survival, birth, death, growth, movement, feeding, sex). Some of these programs are called *cellular automations (CA)*; others are called *genetic algorithms*.

CA programs have been used to accurately model biological organisms and study the spread of communicable diseases like AIDs. CA programs have also been used in evolution, ant colonies, bee colonies, and a host of other uses. Chaos algorithms can be added into the programs to generate randomness. One interesting use of CA programs is to optimize neural networks by creating and wiring neural network layers.

Genetic algorithms (GA) evolve in a Darwinian fashion—survival of the fittest. Two GA programs can mate and mix their binary code. If the offspring is as healthy or has greater health than its parents, it will likely survive.

Are these programs alive? It depends upon your definition of life. What happens when these programs are encased in and control mobile robots? What if the programs evolve and develop higher levels of programming? How about if the robots can eventually self-replicate?

Nanotechnology—Are we alive yet?

Nanotechnology enables the creation of machines the size of microbes. IBM is making progress in manipulating atoms and molecules to create simple machines and electronics (transistors and wire). So far, there appears to be no restriction on how small one can make an object. Bacteria-sized robots are theoretically possible. Again, if tiny robots are created that can gather raw materials to self-replicate, would we consider these robots alive?

Some scientists predict that silicon life will be the next evolutionary step, replacing carbon life forms on this planet. What we call electronics and robotics will evolve into self-creating, self-replicating silicon life.

Whether or not silicon life is or will become the next major evolutionary step on Earth will not be debated. The focus of this chapter remains on the development of artificial intelligence (consciousness) and artificial life.

A little history

The progression of computer technology over the last five decades is staggering. In 1946 the ENIAC computer filled a large area with electronic equipment. The computer was almost 100 feet long, 8 feet high, 3 feet deep, and weighed 30 tons. ENIAC contained 18,000 tubes, 70,000 resistors, 10,000 capacitors, 6,000 switches, and 1,500 electromagnetic relays. ENIAC could perform 5,000 additions per second, 357 multiplications per second, and up to 38 divisions per second. Today that same 1946 computer could be condensed on a tiny piece of silicon less than $\frac{1}{4}$ inch square.

Physicist Robert Jastrow stated in 1981 in *The Enchanted Loom* that, "The first generation of computers was a billion times clumsier and less efficient than the human brain. Today, the gap has narrowed a thousandfold."

Science is progressing unrelentingly toward creating artificial intelligence. This is something we may see in our lifetime. From the standpoint of creating competent AI, it's a small step to generating superior intelligence in machines. That's a dream, many scientists will tell you, trying to retain the waxing illusion that human intelligence is and forever will be unsurpassed. I don't take any comfort in that illusion. AI is an evolving, uncompromising, unrelenting reality.

Greater than us

Would we as the human race produce an intelligence superior to our own? If you think about it, in the long run we may need to just to survive. Think of the advantages for the first government that produced an AI with an IQ of 300. It could be set to work on the national economy, pollution, military strategy, medical and scientific research, and, of course, designing still smarter machines and androids. It's possible that the next theory of the universe will not be put forth by a human (Albert Einstein) but by competent AI.

The locked cage

Why is a superior intelligence so important? This is best illustrated with a story. I once heard or read this story, and I'm afraid that I don't remember the author of the story and to him or her I apologize. And if I changed the story a bit in the retelling, I apoligize for that also.

Ten chimpanzees are locked in a cage. The cage door has a lock on it. The lock requires an intelligence quotient (IQ) of approximately 90 to reason how to unlock the lock and open the cage door. Each chimp in the cage has been tested, and they all have an IQ of about 60. Could the 10 chimps working together find a way to unlock the cage door? The answer is a NO! Intelligence is not cumulative. If it were, the 10 chimps working together could equal an IQ of 600, more than enough to open the cage door. In real life, intelligence is not cumulative, and the chimps would remain caged.

In the real world we have problems like pollution, weak economies, cancer, AIDs, the quest for longevity—any and all facets of scientific research that can be substituted for the lock on the cage door. The importance of generating superior artificial intelligence becomes clearly apparent. AI may be able to uncover the keys to solving these problems while the answers remain hidden to us.

I don't believe that this potential of superior AI is being overlooked by the governments of the world. It's quite possible that the next "Manhattan Project" undertaken by (hopefully) this country will be for creating superior artificial intelligence.

We as a race will take no comfort in appearing as intelligent as a chimpanzee to a machine. Science fiction writers have long written on competent AI's running amok. For instance, the computer HAL in Arthur C. Clarke's *2001*, *Colossus the Corbin Project*, and the main computer in *Terminator I* and *Terminator II*.

So to the future AI programmers reading this text, I have a message: "Don't forget that off switch!"

Biotechnology

Advances in biotechnology will soon allow us to alter our own genetics. With this power it becomes possible to enhance our brain to increase our own intelligence. While possible, it opens up the human race to unforeseeable repercussions of gene altering in subsequent generations that may be catastrophic. This makes generating superior intelligence in machines much safer, at least for the time being.

Hype versus reality

Neural networks have been overhyped since their inception. So it's easy to dismiss my remarks concerning AI and neural networks as just more hyperbole, which people have been saying for years. And it is true that people have predicted the emergence of human-like intelligence in machines. If progress continues as rapidly as it has in the past 20 years, I feel that human-level intelligence in machines will be competent within 50 years.

What are neural networks?

I have mentioned neural networks, but let's discuss them now in more detail. Neural networks are artificial systems (hardware and software) that function and learn based upon models derived from studying the biological systems of the brain. Networks may be implemented in either software/operating systems or hardware. In mimicking the biological systems in the brain, neural networks have taken strides in realizing the potential of artificial intelligence in computers, such as machine vision, voice recognition, and speech.

Neural networks can perform visual recognition. They can learn to read or perform quality control by visual analysis of parts. Other networks can be taught to respond to verbal commands (speech recognition) and generate speech. Statistical nets can predict the future behavior or probability of complex nonlinear systems based on historical examples. These networks have been used to predict oil prices, monitor aircraft electronics, and forecast the weather. Neural networks have also successfully been employed to evaluate the stock market, mortgage loan applicants, and life insurance contracts better than standard rule-based expert-system programs.

What is artificial intelligence?

This is a legitimate question. We most certainly will develop neural networks that are intelligent before we develop nets that are or become conscious. So in attempting to create neural networks that are intelligent or demonstrate intelligence, what criteria should one use to determine if this goal has been achieved?

Alan Turing, a British mathematician, devised an interesting procedural test that is generally accepted to determine machine intelligence. Turing proposed that the machine's intelligence be tested using a person. The person and the machine are connected to one another via a teletype. Through the teletype, the pair would type messages to one another and conduct a conversation. If the machine could simulate intelligence well enough to carry on a conversation without the person being able to determine whether a machine or person existed at the other teletype, the machine could be classified as intelligent. This is called the Turing test and is one criterion used to determine artificial intelligence.

Although the Turing test is well accepted, it isn't a definitive test for artificial intelligence. There are a number of "completely dumb" language processing programs that come close to passing the Turing test. The most famous program is named ELIZA developed by Joseph Weizenbaum at the Massachusetts Institute of Technology. ELIZA simulates a psychologist, and you are able to conduct a conversation with ELIZA. For instance, if you typed to ELIZA that you missed your father, ELIZA may respond with, "Why do you miss your father?" or, "Tell me more about your father." These responses may lead you to believe that ELIZA understands what you have said. It doesn't. The responses are clever programming tricks constructed from your statements.

Therefore, if we like, we could do away with the Turing test and consider a different criterion. Perhaps consciousness or self-awareness would be a better signpost of intelligence. A self-aware machine would certainly know that it is intelligent. Another criterion, one that is more simple and direct and is used in this book, is the ability to learn from experience.

Of course, we could abandon logical approximations and state that intelligence is achieved in systems that develop a sense of humor. As far as I know, humans are the only animals that laugh, and perhaps humor and emotion will end up being the truest test of all.

Using neural networks in robots

So how do neural networks help our robotics work today? Well, we're a way off from creating competent AI, let alone putting it into one of our robots. But neural technology can control robotic function, and in many cases it can perform superior to standard CPU control and programming. By using neural networks in our robots, they can perform small operational miracles without using a standard computer, a CPU, or programming. In Chapter 7 we will design two neuron systems that can track a light source. Place this system on a mobile robot, and it will follow a light source anywhere. Chapter 8 discusses BEAM robotics and Mark Tilden, who designs transistor networks (nervous networks) that allow legged robots to walk and perform other functions.

Another neural process that is making great strides is called *subsumption architecture,* which uses layered stimulus-response.

Subsumption architecture

Subsumption architecture, developed by Dr. Rodney Brooks at MIT, illustrates that relatively simple stimulus-response neural systems, when placed in robotics, can develop high-level, complex behaviors. Subsumption architecture is covered more fully in Chapter 8.

Tiny nets

Small neural network programs can also be written in microcontrollers like the Basic Stamp 1 (BS-1) and BS-2. For more information on these microcontrollers, see Chapter 6.

Power

3

ROBOTS NEED POWER TO FUNCTION, AND MOST ROBOTS use electrical power. The two main sources of untethered electrical power for mobile robotics are batteries and solar cells.

Solar cells

Solar cells produce electrical power from sunlight. If enough solar cells are strung together, sufficient power can be generated to operate a robot directly. Usually, solar robots are designed as small as possible while still being able to perform their designated functions. Solar robots are constructed using high-strength, lightweight materials and low-power electronics.

Reducing or keeping a robot's weight down reduces the electrical power needed for movement and locomotion. Weight reduction is a good design goal for battery-powered and rechargeable robots. Lightweight robots will be able to operate longer on a given power supply than a heavier counterpart.

Another method of using solar cells doesn't drive the robot directly. Instead, solar cells are used as a power source for recharging batteries. This hybrid power supply reduces the required capacity of solar cells needed to operate the robot directly. However, the robot can only function a percentage of the time that it spends recharging its power supply.

A third method of using solar cells combines the technologies of the first two methods. Here we build what is commonly called a solar engine. The circuit is simple in function. The main components are a solar cell, main capacitor, and a slow oscillating or trigger circuit. When exposed to light, the solar cell begins charging a large capacitor. The solar cell/capacitor provides electrical power to the rest of the circuit. As the voltage on the capacitor rises, a slow oscillating circuit begins pulsing. When sufficient power has been stored in the capacitor, the oscillating circuit pulse triggers an SCR. The SCR dumps the stored power in the capacitor for robotic function, usually through the robot's drive system. The cycle then repeats. The solar engine may be used in a variety of innovative robotic designs.

Building a solar engine

The solar engine is commonly used as an onboard power plant for BEAM-type robots, sometimes called living robots. The inspiration for this solar engine came originally from Dave Hrynkiw from Canada, who uses a similar design to power a solar ball robot. I liked the electrical function so much that I decided to design my own solar engine. In doing so I was able to create a new circuit that improves the efficiency of the original design.

Figure 3.1 is the schematic for the solar engine. Here's how it works. The solar cell charges the main 4,700-uF capacitor. As the capacitor charges, voltage level of the circuit increases. The UJT begins oscillating and sending a trigger pulse to the SCR. When the circuit voltage has risen to about 2 volts from the main capacitor, the trigger pulse is sufficient to turn on the SCR. When the SCR turns on, all the store power in the main capacitor is

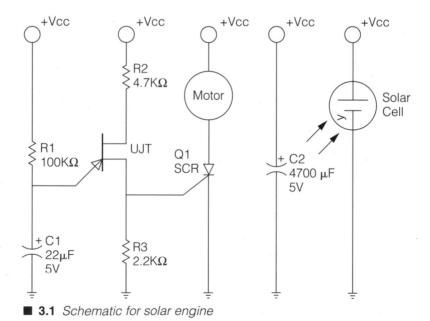

■ **3.1** *Schematic for solar engine*

24

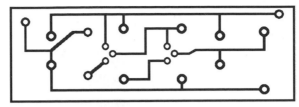

■ **3.2** *PCB foil pattern*

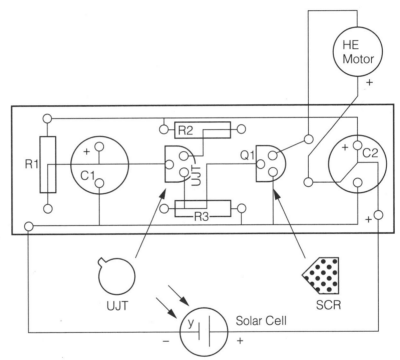

■ **3.3** *Parts placement on PCB*

dumped through the high-efficiency (HE) motor. The motor spins momentarily as the capacitor discharges, then stops. The cycle repeats.

The solar engine circuit is simple and noncrucial. It may be constructed using point-to-point wiring on a prototyping breadboard. PCB pattern is shown in Fig. 3.2 for those who want to make the printed circuit board. The solar engine kit, see parts list, has a PC board included. Figure 3.3 shows the PCB parts placement. The complete solar engine is shown in Fig. 3.4.

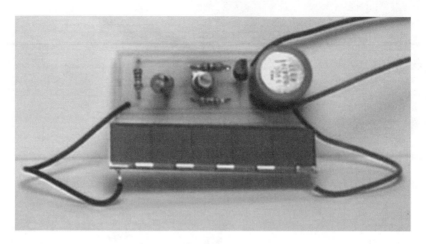

■ **3.4** *Complete solar engine*

High-efficiency motor

Not all electrical motors are high efficiency (HE). For instance, the small electrical motors sold at your local Radio Shack are low-efficiency type. There is a simple way to determine if a motor is an HE type. Spin the rotor of the motor. If it spins smoothly and continues to spin momentarily when it's released, then it's probably an HE type. If when you spin the rotor it feels clunky or there is resistance, it probably is a low-efficiency type.

Caveats

The solar cell used in this circuit is high voltage, high efficiency. Typically, solar cells supply approximately .5 volts at various currents, depending on the size of the cell. The solar cell used in this circuit is rated at 1.5V, but I've seen it charge a capacitor up to 2.5V under no-load conditions.

I'm sure some people planning to build this circuit are already thinking about adding a few more solar cells to speed charging. While one can stack low-voltage solar cells in series to reach a usable (2–2.5V) voltage, one should not add solar cells to increase the current. There's a reason for this. In order for the circuit to recycle, the current through the SCR must stop (or at least be very minimal) for the SCR to close. If there is too much current supplied by the solar cell(s), the SCR will stay turned on. If this happens, the circuit will stop cycling.

The solar cell used in the circuit is balanced for proper operation. One component you may want to change is the main capacitor. You can use one of the supercaps available to store more electrical power, but be aware that when using these capacitors it will take that much longer to cycle.

Uses

The circuit may be used in many novel and innovative ways. It may be used as an onboard power plant for a solar racer, to supply power to an HE motor for robotic locomotion (see the Chapter 8 BEAM solaroller project), audio graffiti, flashing LEDs, or as the demo circuit pictured in Fig. 3.5 spinning an American Flag. The attractiveness of the circuit is that it operates perpetually, or at least until one of the components breaks, which means it should operate for years.

Parts List

- ☐ (1) 2N2646
- ☐ (1) 2N5060
- ☐ (1) 22-uF cap
- ☐ (1) 47000-uF cap
- ☐ (1) HE motor

■ **3.5** *Solar engine flag spinner*

☐ (1) Solar cell
☐ (1) PCB
☐ R1 100K ¼ watt
☐ R2 4.7K ¼ watt
☐ R3 2.2K ¼ watt

The kit contains all the above parts for $25.50 post paid. It is available from:

Images Company
PO Box 140742
Staten Island, NY 10314
(718) 698-8305

Batteries

Batteries are by far the most commonly used electrical power supply for robotics. Batteries are so commonplace that it's easy to take them for granted, but understanding batteries will help you choose ones that will optimize your robot's design. The rest of this chapter examines batteries.

There are hundreds of different batteries, but we will look at the most common batteries used by hobbyists: carbon-zinc, alkaline, nickel-cadmium, lead-acid, and lithium.

Battery power

Regardless of battery type, battery power is measured in amp-hours, which is the current (measured in amps or milliamps) multiplied by the time (hours) that current is flowing from the battery. What does that mean to us? Well, it's pretty straightforward. Suppose a battery is rated at 2 amp hours. This means the battery can supply 2 amps of current for 1 hour. If the current draw from the battery is only 1 amp, the battery will last 2 hours. If the current is further reduced to 500 mA, the battery will last 4 hours.

Your mileage may vary. Batteries provide more access to their electrical power when used intermittently, allowing time for the battery to recuperate. Continuous duty is efficient if the load is light. For robotics, especially powering drive motors and components, we often don't have these options. In this case one tries to provide greater battery capacity.

Battery voltage

Battery voltage varies through the life of a battery. If you measure the voltage on a fresh "D"-sized 1.5-volt alkaline battery, it will read approximately 1.65 volts. As the battery discharges, its voltage drops. The battery is considered dead when the voltage drops to around 1.0 volts. Typical discharge curves for carbon-zinc, alkaline, and nicad (NiCd) batteries are illustrated in Fig. 3.6.

Notice that a fresh NiCd 1.5V battery actually delivers about 1.35 volts. While its initial voltage is lower, its discharge curve is fairly flat compared to carbon-zinc and alkaline delivering a constant 1.2 volts.

Primary batteries

Primary batteries are one-time-use batteries. The batteries we will look at in this class deliver 1.5 volts per cell. They are designed to deliver their rated electrical capacity and then be discarded. When building robotic systems, discarding depleted primary batteries can become expensive, especially if the robotic system is to be used often.

One advantage to using primary batteries is that they typically have a greater electrical capacity then rechargeables. If one is engaged in a function (i.e., robotic war) that requires the highest power density available for one-shot use, primary batteries may be the way to go.

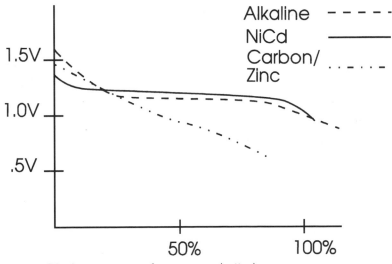

■ **3.6** *Discharge curves for common batteries*

Secondary batteries

Secondary batteries are rechargeable. The most common rechargeable batteries are nickel-cadmium (nicads) and lead acid. We will start with nicads.

One disadvantage to NiCd batteries is a lower voltage, 1.2 volts per cell. So a "C" cell battery will deliver about 1.2 volts instead of 1.5 volts. The effect becomes more pronounced when using multiple cells. For instance, a "9V" NiCd battery will deliver approximately 7.2 volts.

Automotive lead-acid batteries are rechargeable but are not really suitable for robotics, the reason being that they are not designed to be completely discharged before recharging. Complete discharge before recharging is called deep cycle. Because of the solar power industry, there are deep-cycle lead-acid batteries available, but you will find that these batteries carry a higher price tag. When building robotic systems you want deep cycle rechargeable batteries.

Secondary batteries, while initially more expensive, are cheaper in the long run. Typically, secondary batteries can be recharged 200 to 1,000 times. In many cases a simple recharging circuit can be built into the robot so that it becomes unnecessary to remove batteries for charging.

Rating primary batteries

As you may have guessed, there are a number of primary batteries available. The differences in batteries relate to the chemistry used in the battery to produce electricity. The choice of a primary battery is a trade-off of price versus energy density, shelf life, temperature range, discharge slope, and peak current capacity.

Carbon-zinc At the low end of primary batteries is the carbon-zinc battery. This battery hasn't changed much since 1868 when it was developed by George Leclanche. Carbon batteries have a low energy density (1–2 WH/cubic inch), poor high-current performance, sloping discharge curve, and bad low-temperature performance. Carbon-zinc batteries are inexpensive but obsolete.

Alkaline-manganese Sold as simply alkaline batteries, the energy density is 2–3 WH/cubic inch. It has improved the low-temperature performance and sloping discharge curve that is not as severe as carbon-zinc batteries. The cost is moderate.

Lithium This is a premium battery with high energy density (8 WH/cubic inch). It has excellent low-temperature and high-temperature performance, a long shelf life (+5 years), and is lightweight. The cost is expensive.

Rating secondary batteries

Nicads and sealed lead-acid batteries are the most common rechargeables. Nicads being the more popular of the two. Both types of batteries have lower energy densities than primary batteries.

Nicads only provide 1.2 volts per cell, in comparison to 1.5 volts per cell of the primary batteries. Manufacturers claim that nicads are good for 200–1000 charge/recharge cycles. However, nicads will die fast if they aren't recharged properly. The life expectancy of nicads is 2–4 years. A fully charged nicad battery will lose its charge in 30–60 days without use.

Nicads are designed to be recharged at 10% of their rated capacity, meaning that if a particular nicad battery is rated at 1 ampere-hour (1 Ah), it is safe to recharge the battery at 100 mA. (1 amp/10 = 100 mA). The terminology used to describe the above-recommended recharge rate is "C/10."

Nicads are designed to be charged using a constant current at the C/10 rates. Because of inefficiencies, it is necessary to charge the battery for 14 hours to get a full charge. While manufacturers claim that it is OK to overcharge NiCd batteries at the C/10 rate, most engineers recommend switching over to a trickle charge after the initial 14 hours at C/10. A trickle charge is usually rated at C/30 or (1/30) of the battery's capacity. A trickle charge for our 1-Ah battery would be around 33 mA (1 amp/30 = 33.3 mA).

Memory Effect

A disadvantage to nicads is the memory effect. If one repeatedly recharges a NiCd battery before it has completely discharged, the batteries form a memory at that recharge level. It then becomes difficult to discharge the battery past that remembered level. Obviously, this can severely limit the battery's capacity. To correct that problem, the battery must be completely discharged by leaving a load connect to the battery for several hours. Once the battery is complete discharged, it can be charged normally and will function properly.

Lead acid

Gelled-electrolyte battery cells (gel-cells) are similar to automotive batteries. They are sealed, maintenance-free, lead-acid batteries. They don't make gel-cells in the familiar D, C, AA, AAA, or 9V battery cases. Gel-cells are typically larger and may be used in larger robots.

Gel-cells are available in numerous voltage ratings, from 2 to 24 volts and current capacities. These batteries may be charged with a current-limited constant voltage or a constant current like nicads. Typically, to charge a gel-cell, one applies a fixed 2.3V–2.6V per cell. Initially the battery will draw a high current that tapers down as it charges. When fully charged, the battery need only draw just a trickle charge (approx. C/500) to maintain itself in a fully charged state.

Gel-cell batteries vary from manufacturer to manufacturer. To safely recharge a gel-cell, you should check the manufacturer's recommendation. In general, a simple charging device can be made using an LM317 voltage regulator. A fixed voltage (2.3V per cell), constant current C/10 is applied to the battery. When the battery reaches a full charge, the constant current source is removed and a regulated voltage is applied.

Many gel-cell batteries do not like to be deep cycled. Therefore, it becomes necessary to monitor battery voltage under load. When battery voltage drops by a specified amount (check manufacturer's data sheet), the battery needs to be charged.

In general

Most robotists use alkaline batteries when a primary battery is called for and nicads when secondary batteries are needed.

NiCd battery charger

NiCd battery chargers are inexpensive. Typically, it is not worth the time and effort to build a stand-alone charger for common-size batteries: AAA, AA, C, D, and 9V. However, most inexpensive chargers will charge only at the C/10 rate, even after the batteries have received a full charge after 14 hours.

The charger we will build will drop the current down to a C/30 rate after the batteries are fully charged. This is the recommended procedure for charging NiCd batteries. This will help provide a long service life to the battery.

In addition, if you want to design a system for charging a particular battery pack or to charge batteries without removing batteries from a robotic project, then a custom NiCd charger is the way to go.

The charger unit itself is still separate from the robot. (Why carry extra weight and components around when you don't have to?) The robot should have a power socket that connects to the power supply. In between the socket and power supply, one should add a double-pole, double-throw switch (DPDT). The DPDT switch connects the power supply to either the robot's circuitry or to the charger. This prevents powering the robots and reducing the current flow to the batteries while the batteries are being charged (see Fig. 3.7).

The power for the charger is supplied by either a standard transformer or a Vdc plug-in wall transformer. I would choose a wall transformer because it supplies a dc voltage. If you use a standard transformer, you must build the power supply using a line cord, switch, fuse, bridge rectifier, and smoothing capacitor.

In either case you should match the transformer (or wall transformer) power output to the battery pack you are charging. Matching the voltage and current to the battery pack reduces the power the LM317 must dissipate. For instance, you don't want to use a 12-volt transformer to charge a 6-volt battery pack.

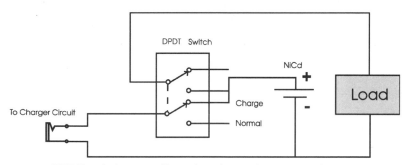

■ **3.7** *DPDT switch controlling charging to battery pack*

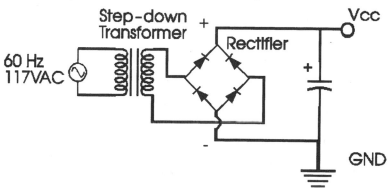

■ **3.8** *Basic power supply for charger circuit*

Figure 3.8 is a basic Vdc power supply for the charger. The power supply can be made to provide either 6V, 12V, 18V, 24V, or 36V depending upon the transformer, bridge rectifier, and capacitor chosen.

The charger circuit is illustrated in Fig. 3.9. It uses an LM317 voltage regulator and a current-limiting resistor. The current-limiting resistor is calculated depending on the current needed to charge the battery.

Current-limiting resistor

Most NiCd battery manufacturers recommend charging the battery at 1/10 of its rated capacity, referred to as C/10. So if an "AA" battery is rated at .850 ampere hours (Ah), it should be charged at 1/10 that capacity or 85 mA for 14 hours. After the batteries are fully charged, manufacturers recommend dropping the current to around C/30 (1/30 of battery capacity) to keep them fully charged without overcharging or damaging the batteries in any way.

For our example, we will configure the charger to recharge four "C" cells in series. Each "C" cell is rated at 2000 mA, so our C/10 rate is 200 mA. The typical voltage rating from this battery is approximately 1.3 volts (4 × 1.3V = 5.2V). We can use a 6-volt transformer with at least a 200 mA output.

To calculate the current-limiting resistor, use the formula:

$$R = 1.25/Icc$$

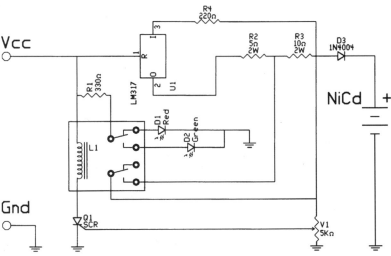

■ **3.9** *Schematic of charger circuit*

where Icc is the desired current. Plugging in our 200 mA (0.2 amp) yields:

$$1.25/.2 = 6.25 \text{ ohms}$$

The current-limiting resistor for this charger should be around 6.25 ohms. In the schematic, this resistor is labeled R2. Notice that the R2 value listed in the schematic is 5 ohms. You should choose a common resistor value as close as possible to the calculated value.

C/30 resistor

To drop the current to a C/30 range, we add another resistor whose value is 2R or about 12.5 ohms. In the schematic (Fig. 3.9), this resistor is labeled R3.

Again, a resistor with the closest value to the calculated value is used. In this case, the value is 10 ohms.

How the charger works

The charger uses an LM317 voltage regulator as a constant current source. The C/10 current-limiting resistor is identified as R2 in Fig. 3.9.

R2, you will notice, is only 5 ohms, as compared to the calculated 6.25 ohms. This standard value is close enough to the calculated value for proper operation. The C/30 resistor is R3 on the schematic. Again, the standard value of 10 ohms is close enough to the calculated value for proper operation. Later on, we will see that it's possible to quickly charge the batteries because of the voltage-sensing capacity of the circuit.

V1 is a 5K potentiometer. It is set to trigger the SCR when the NiCd batteries are fully charged. The SCR, once triggered, allows current to flow through a DPDT relay.

When power is applied to the circuit, current flows through the LM317, charging the batteries at a C/10 rate. Resistor R3 is shorted by one-half of the DPDT relay. Current also flows through resistor R1, which is a current-limiting resistor for LEDs D1 and D2. Upon power-up, the red LED D1 will be lit. When the red LED is lit, it means the circuit is charging.

As the batteries charge, the voltage drop across V1 begins to climb. After about 14 hours, the voltage drop across V1 is great enough to trigger the SCR. When the SCR is triggered, current flows through the coil of the DPDT relay. The relay switches, causing the red LED to go out and the green LED to turn on. The green LED signals that the batteries are fully charged. The other half of the relay switches, opening up the short on resistor R3. With R3 now in the current path, the current flowing to the NiCd batteries is cut to a C/30 level. Diode D3 prevents any current from the batteries flowing back into the circuit.

Determining the trigger voltage from V1

For the circuit to function properly, the SCR must turn on when the NiCd batteries are fully charged. The easiest (best) way to do this is to place depleted batteries in the charger, charge the batteries for 14 hours, then adjust V1. With fully charged batteries, slowly turn V1 until the relay clicks and the green LED turns on.

Design notes

When building a charger for your application, keep these points in mind. The main considerations are choosing the C/10 and C/30 current-limiting resistors. Use the formulas described above for selecting these values. Current-limiting resistors should be rated around 2 watts.

If charging current is high (greater than 250 mA), heat-sink the LM317. If the charger switched on without the NiCds connected, the relay will switch immediately, turning on the green LED and providing a C/30 current.

When building a charger for higher voltages, increase the valve of R1 proportionally to limit the current flowing through the LEDs. For instance, for a 12V unit, make R1 680 ohms; for a 24V unit, make R1 1.2K ohms.

At high voltages you may need a low-ohm-value, current-limiting resistor connected to the DPDT relay. Measure the current C/10 and C/30 current flowing to the batteries. These measurements will ensure that the proper current is being supplied to the batteries.

Series/parallel charging

How the batteries are configured determines the voltage and current of the transformer one should use. If you have eight "C" battery cells in parallel, you need to multiple the current requirements of each individual cell by eight. If the cell is rated at 1200 mAh, the C/10 requirement per battery is 120 mA. For eight cells in parallel (8×120 mA = 960 mA), you need close to one amp (.96 amp) of current. The voltage required is just 1.5 volts. The ideal transformer's output would be 1.5V at 1 amp. If the eight cells were held in series, the current requirements would be 120 mA at 12 volts.

Fast charger

Many of today's NiCds are capable of accepting a fast charge, provided that the circuit can sense when the batteries are fully charged and drop the current to C/30. Typically, to fast-charge a battery, you double the current for half the time, so you charge a battery at C/5 for 7 hours.

Although I haven't tried the above circuit for fast charging, there is no reason why it wouldn't work. You may want to start with a C/10 charging current and adjust V1, then switch out resistor R2 for one with half the valve.

Parts list

- [ ] U1 LM 317 voltage regulator
- [ ] L1 DPDT relay (5V or 12V)
- [ ] D1 red LED
- [ ] D2 green LED
- [ ] D3 1N4004
- [ ] Q1 SCR
- [ ] V1 5K PC-mounted potentiometer
- [ ] R1 330-ohm, $\frac{1}{4}$ watt
- [ ] R2 5-ohm, 2 watt
- [ ] R3 10 ohm, 2 watt
- [ ] R4 220 ohm, $\frac{1}{4}$ watt
- [ ] Wall transformer (see "NiCd battery charger" earlier in this chapter.

Solar-powered battery charger

Once you have designed a battery charger for a rechargable battery pack, you can convert it to a solar-powered battery charger. You need to replace the step-down transformer (or wall transformer) with a combination of solar cells that will equal the power delivered by the transformer. Points to keep in mind when planning a solar power system are:

- ☐ The average illumination received by the solar panel.
- ☐ The hours of illumination needed to recharge power supply versus work period.

Movement and drive systems

4

This chapter provides an overview of movement and drive components that can be used in robots. Most of the components discussed in this chapter either have sample circuits contained in this chapter or are used in robots elsewhere in this book.

Here is a list of components we will work with: air muscles, nitinol wire, stepper motors, gear dc motors, servo motors, and solenoids.

Air muscle

The air muscle (see Fig. 4.1) is a type of pneumatic device that produces linear motion with the application of pressurized air. Much like a human muscle, it contracts when activated. You may think, Well this is nothing new; pneumatic cylinders have been around for decades and do the same thing. However, the air muscle is an innovation: lower in cost, lightweight, flexible, and much easier (safer) to use.

The air muscle boasts a power-to-weight ratio of 400:1. Because most of its components are soft plastic and rubber, the air muscle can work when it's wet or even under water. The flexible nature of the products allows it to function when bent around curved surfaces. So, as you can see, the air muscle has a number of features that surpass standard pneumatic cylinders. The air muscle is described completely and used in Chapter 16 to make an anthorobotic (android) hand.

Nitinol wire

Nitinol is a metal that belongs to a class of materials called shaped memory alloys (SMA). Nitinol is commonly sold in wire form. When heated, the material can contract up to 10% of its length. The contraction of the material produces linear motion. In addition to the

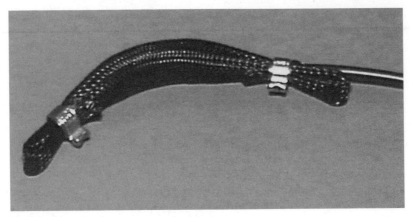

■ **4.1** *Air muscle*

contraction property, the material also exhibits a shaped memory effect (SME).

The SME is a unique property of this alloy. When heated up to its crucial (transition) temperature, the material returns to a predefined shape. The predefined shape is one the material is trained (heat annealed) to remember. The material is formed into the training shape. The material is then forcibly confined to the training shape as the material is annealed (heated) above the transitional temperature. This realigns the crystalline structure to the shape. Now the object will return to this shape whenever it's heated to its transition temperature. So a trained object could be twisted and folded out of shape, then heated to return the object back to its original shape.

These unique properties of SMEs rely on the crystalline structure of the material. The shape-resuming force approaches 22,000 lbs per square inch. It's very unlikely that anyone will be working with such large cross sections of material. Even thin wires of the material produce an impressive force. For instance, a 6-mil wire generates a contractive force of 11 ounces.

When nitinol wire contacts up to 10% of its overall length, its volume remains the same. As the wire contracts, its diameter increases proportionally, keeping the net volume of the wire constant.

The easiest way to heat nitinol wire is by passing a steady electric dc current through it (see Fig. 4.2). Using a steady dc current for an extended period of time can damage the wire due to uneven ohmic heating.

Proportional control and steady-state contraction of the material (without damage) can be achieved using a pulse width modulation circuit (PWM) to supply the electrical current.

Some robotists have used nitinol wire to create a motorless walking robot. While the robot can walk, it does so exceedingly slowly due to the time required for cycling (heating-cooling) of the nitinol material. This walker robot is extremely lightweight (a few ounces at most) and is not powerful enough to carry its own power supply.

To learn more of the capabilities of this remarkable material, let's look at other applications that emphasize the contraction ability of the material. Figure 4.3 shows a mechanical butterfly. Nitinol wire adds movement to the wings, and the butterfly may be connected to a solar engine (see Chapter 3) for power to create an interesting robotic application. Figure 4.4 shows a rocker ball demonstration device. The nitinol actuator operates about 20,000 cycles per day and will last for years.

Nitinol wire loops can be used to produce rotary motion. Figure 4.5 illustrates a simple heat engine. The nitinol loop is guided by a groove in each wheel, and the smaller wheel is made of brass for good heat conduction. By placing the smaller wheel in hot water, the wheels begin to spin. The heat engine can also function using solar energy. Focusing sunlight from a 3" magnifying glass onto the brass wheel will also activate the engine.

Nitinol can be used to physically close mechanical push-button switches, as an actuator in production of lightweight air valves, and many other linear motion applications.

Solenoids

Solenoids are electromechanical devices. A typical solenoid consists of a coil of wire that has a metal plunger through its center.

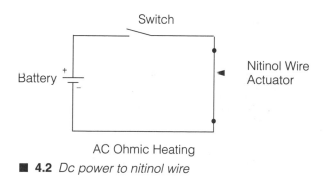

■ **4.2** *Dc power to nitinol wire*

■ **4.3** *Nitinol butterfly*

■ **4.4** *Rocker ball demo*

When energized, the coil creates a magnetic field that either pulls or pushes the metal plunger (see Fig. 4.6).

Rotary solenoids

A *rotary solenoid* is a derivative of the standard solenoid (see Fig. 4.7). Instead of producing a linear motion, it produces a rotary motion. A rotary solenoid is used to create a robotic fish (see Chapter 13).

■ **4.5** *Heat engine*

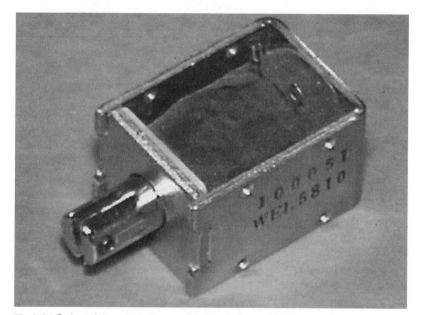

■ **4.6** *Solenoid*

■ **4.7** *Rotary solenoid*

Movement and drive systems

Stepper motors

Stepper motors may be used for locomotion, movement, steering, and positioning control. These motors are an integrated component in many commercial and industrial computer-controlled equipment. For home PC users, stepper motors can be found in disk drives and printers.

Stepper motors are unique because they can be controlled using digital circuits. They are capable of precise incremental shaft rotation. This makes stepper motors ideal for rotary or linear positioning.

Because stepper motors are widely used in industry, they come in a variety of shapes, sizes, and specifications (see Fig. 4.8A).

When power is applied to a standard electric motor, the rotor begins turning smoothly. Speed and position of the motor's rotor are a function of voltage, load on the motor, and time. Precise positioning of the rotor is not possible.

A stepper motor, however, runs on a sequence of electric pulses to the windings of the motor. Each pulse to a winding turns the rotor by a precise, predetermined amount. The incremental movement of the rotor are often called *steps*. Hence the name *stepper motors*.

■ **4.8A** *Stepper motor*

Not all stepper motors rotate the shaft (rotor) by the same amount per step. They are manufactured with different degrees of rotation per step (or pulse). The optimum degree per step will depend on the particular application. Stepper motor specifications clearly state the degree of rotation per step. You can find a variety of stepper motors whose range of rotation per step can vary from a fraction of a degree (i.e., .72 degrees) to many degrees (i.e., 22.5 degrees).

Stepper motor circuit

Figure 4.8B is a schematic of a stepper motor driver circuit. The stepper motor in the circuit is a unipolar (six-wire) type. IC U1 is a 555 timer that is set up in a stable mode to output square wave-clocking pulses on pin 3. U2 is a stepper motor controlling chip UNC5804. The clocking pulses received on pin 11 of the UCN 5804 turn the stepper motor. Each pulse received on pin 11 turns the stepper motor one step. The faster the clocking pulses, the quicker the stepper motor turns.

In this sample circuit, the clocking pulses are produced by a 555 timer. Clocking pulses can be generated by any number of sources, like a microcontroller (discussed in Chapter 7) or a photoresistive neuron (discussed in Chapter 5). Switch SW1 is a fast/slow control. SW2 controls the stepper motor direction. Stepper motors are used in making a robotic platform in Chapter 10.

Servo motors

Servo motors are geared dc motors with positional control feedback and are used for position control. The shaft of the motor can be positioned or rotated through 180 degrees. They are commonly used in the hobby R/C market for controlling model cars, airplanes, boats, and helicopters.

Because of their widespread use in the hobby market, servo motors are available in a number of stock sizes (see Fig. 4.9). While larger industrial servos are also available, they are priced out of range for most hobby applications. This book is restricted to the hobby servo motors that are inexpensive and readily available.

There are three leads to a servo. Two are for power, + 4-6 volts and ground.

The third lead feeds a position control signal to the motor, and the control signal is a variable-width pulse. A neutral, midrange positional pulse is a 1.5-ms (millisecond) pulse, which is sent 50 times

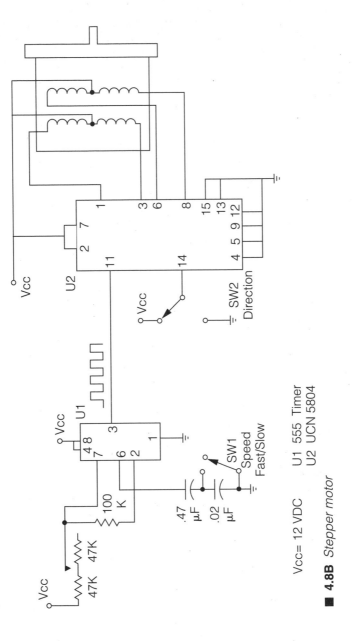

4.8B *Stepper motor*

Vcc= 12 VDC   U1 555 Timer
                U2 UCN 5804

SW1
Speed
Fast/Slow

SW2
Direction

Vcc

U1

U2

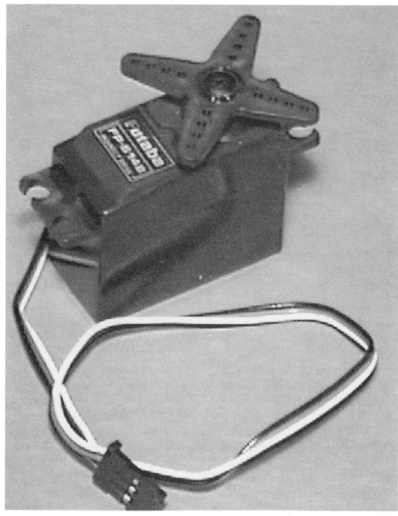

■ **4.9** *Servo motors*

(20 ms) a second to the motor. This pulse signal will cause the shaft to locate itself at the midway position +/–90 degrees.

The shaft rotation on a servo motor is limited to approximately 180 degrees (+/–90 degrees from center position). A 1-ms pulse will rotate the shaft all the way to the left (see Fig. 4.10), while a 2-ms pulse will turn the shaft all the way to the right. By varying the pulse width between 1 and 2 ms, the servo motor shaft can be rotated to any degree position within its range.

You may feel that providing the pulse signal is a complex job; it isn't. The basic stamp covered in Chapter 6 uses only a few lines of code

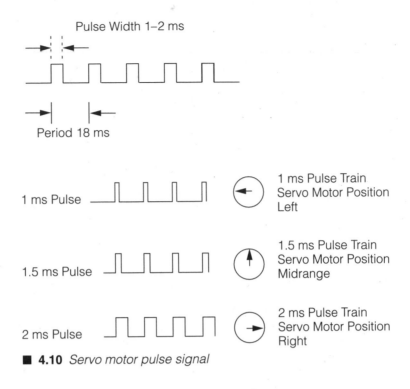

Pulse Width 1–2 ms

Period 18 ms

1 ms Pulse

1 ms Pulse Train
Servo Motor Position
Left

1.5 ms Pulse

1.5 ms Pulse Train
Servo Motor Position
Midrange

2 ms Pulse

2 ms Pulse Train
Servo Motor Position
Right

■ **4.10** *Servo motor pulse signal*

to control a servo motor. And the stamp can control up to eight servo motors at a time. Another viable method is to use the servo control system used in R/C systems. The last alternative is to make your own circuit.

Making a servo circuit isn't as difficult as it may first appear. Figure 4.11 uses a 556 dual timer to control a servo motor. The 556 has two independent timers. To see the function more clearly, see Fig. 4.12. Here two separate 555 timers are used. One timer is set in astable mode. The astable timer outputs a 55-Hz square wave with a 1-ms negative component. The output from this timer is connected to the second 555 timer, which is set up in monostable mode.

The monostable timer outputs a positive pulse from pin 5 each time it receives a negative pulse from timer 1. The positive pulse from timer 2 can be varied from 1 ms to 2 ms using the 10K potentiometer.

You may need to fiddle with the resistance values of R1 and R2 in Fig. 4.11, depending on the servo motor being used. As always, take caution that the servo isn't stalled (pushing against its internal rotational stops).

When working with servo motors, I found that to achieve full rotational movement from the shaft, I needed to use pulses shorter than 1 ms and greater than 2 ms.

As you work with and gain experience using servo motors, you can also run outside the standard pulse widths (shorter and longer pulses) to achieve fuller or complete 180-degree shaft rotation. The standard pulse widths sometimes do not rotate the shaft to each end stop.

Before you do so, you need to understand that if a pulse signal falls outside of the range where the servo shaft can rotate, the servo will fight against its internal rotational stop in an effort to rotate the shaft to the position called for by the pulse. The stalled motor will increase the current to the motor and wear on the gearing inside the motor.

An example may clear this up. Suppose you have a servo motor and you are providing a 2.8-ms pulse to rotate the shaft all the way to the right. As long as the shaft can rotate to that position, you're fine. But suppose a different servo reaches its maximum right rotation with a 2.5-ms pulse. If you start sending a 2.8-ms pulse width, you are ordering the servo to turn further than it physically can.

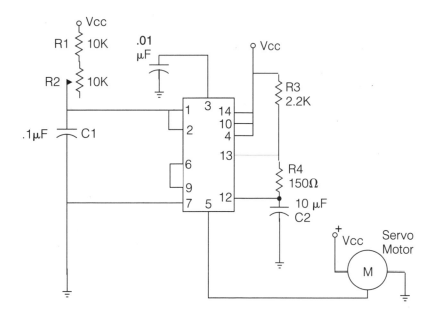

U1 = 556 Dual Timer

■ **4.11** *Servo motor circuit 556*

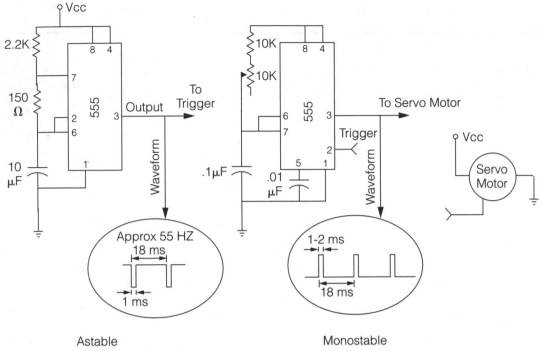

■ **4.12** *Servo motor circuit 555*

Because the servo is fighting against the internal stop, current drain goes up, you increase wear on the gearing, and you may burn out the servo.

This problem usually occurs when the original servo motor tested is changed and the replacement motor doesn't need the modified pulse widths. As a rule of thumb, if using pulse outside of the recommended 1–2-ms pulse, always check to make sure the servo motor isn't stalled in position.

Servo motors are used to make the walker robot in Chapter 11. The walker uses a basic stamp to control the servo motors, and the basic stamp and servo motor application are covered in Chapter 6.

Dc motors

Dc hobby motors can be applied to movement and locomotion (see Fig. 4.13). Most dc motor specifications show a high rpm and low torque. Robotics need low rpm and high torque. Gearboxes can be attached to motors to increase their torque while reducing the rpm (see Fig. 4.14). The gearbox usually specifies a ratio that describes the rpm in to the rpm out. For instance, a dc motor with an rpm of 8000 is connected to a 1000:1 gearbox. What is the output rpm? (8000 rpm/1000 = 8). The torque of the motor is substantially increased. You could estimate that the torque will increase by the same value that the rpm decreased. In reality, no conversion is 100% efficient; there will be efficiency losses. Some dc motors, called gear head motors, are built with a gearbox attached (see Fig. 4.15).

Dc motor H-bridge

When building a robot, one wants to control (turn on or off) the dc motor via a simple circuit or digital signal. In addition, one would also like to be able to reverse the motor's direction. An H-bridge

■ **4.13** *Dc motor*

■ **4.14** *DC motor with gearbox*

fills all these requirements. It should be understood that the dc motor refers to stand-alone dc motors and motors connected to gearboxes or with gear heads.

The H-bridge is made of four transistors. (Some robotists use MOSFET transistors; I use Darlington transistors.) Each transistor acts like a simple switch (see Fig. 4.16a). When switches SW1 and SW4 are closed (see Fig. 4.16b), the motor rotates in one direction. When switches SW2 and SW3 are closed, the motor rotates in the opposite direction.

By using the switches properly, we can reverse the current flow to the motor, which in turn changes the motor's shaft rotation. Figure 4.17 is an H-bridge circuit using transistors. An H-bridge circuit is used in Chapter 5 in the sensor tester robot.

Pulse width modulation

The H-bridge controls the on and off function and direction of dc motors. The function of the H-bridge can be enhanced by using pulse width modulation (PWM) to control the speed of the motor. PWM signal is illustrated in Fig. 4.18. When the PWM signal is high, the motor is on; when the PWM signal is low, the motor is off. Since the signal turns the motor on and off very quickly, the voltage delivered to the motor becomes an average of the time-on versus the time-off period of the cycle (T-on/T-period). The greater the on time, the higher the average voltage. The average voltage is always less than the (Vdc steady-state) voltage delivered (Vcc).

PWM essentials control the motor speed.

Motors are inductive loads. When current is switched on and off, a transient voltage is generated in the (motor) windings that can damage the solid-state components used in the H-bridge. This

■ **4.15** *Dc motor with gearbox head*

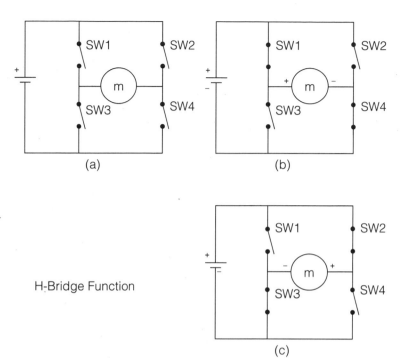

H-Bridge Function

(a)

(b)

(c)

■ **4.16** *H-Bridge using switches*

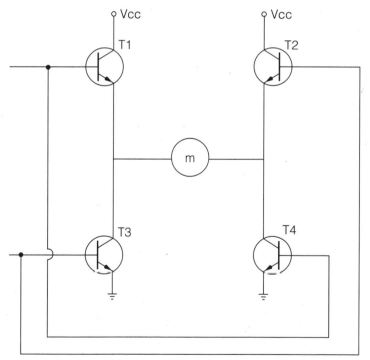

■ **4.17** *H bridge using transistors*

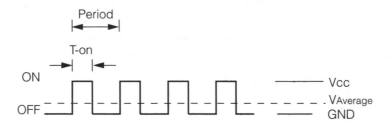

■ **4.18** *Pulse width modulation (PWM) for H-Bridge*

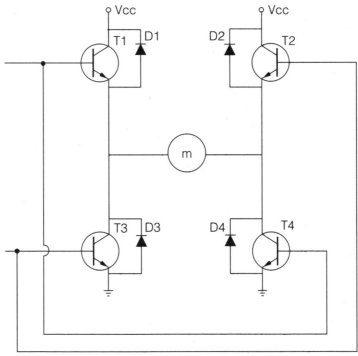

■ **4.19** *Transistor H-Bridge with diode protection*

problem can be avoided by using a snubber-diodes bridge across each transistor, and this is illustrated in Fig. 4.19.

The snubber diodes create a voltage path to ground for the transient voltage. This effectively safeguards the semiconductor the diode is bridged over. The snubber diodes should be rated to handle the normal current the motor typically draws.

Sensors

TYPICALLY, ROBOTIC SENSORS MIMIC BIOLOGICAL SENSES like hearing, sight, touch, smell, and taste. Balance and body position derived from the inner ear are sometimes considered a sixth sense. Biological senses are neurally based, while robotic senses are electrically based. One could argue that they are both electrically based because the neural pathways and signals pass an electrochemical signal. However, neural sensors function differently than electrically based sensors. So, not to confuse technologies, it's important to define robotic sensors as electrically based.

If one wants to truly imitate biological senses, neural sensors are needed. As an example of a neural sensor, let's examine the human ear. The human ear is not a linear instrument. Its response to sound is logarithmic. Because of this, a tenfold increase in sound intensity only results in a perceived doubling of sound. A common sound sensor, in this case a microphone, has a linear response to sound intensity.

Sensors detect and/or measure an aspect of the environment and produce a proportional electric signal. The signal information must then be interpreted by the intelligence (CPU) or neural network on the robot. While we may categorize the sensors as they relate to human senses, sensors are typically divided by the type of energy that the sensor responds to, such as light, sound, and heat. The sensors one incorporates into a robot will depend upon its intended operating environment and application.

Signal conditioning

When planning sensors for a robot, one must decide how the robot will read the sensor signal. Many sensors are resistance-type sensors, meaning that the sensor varies its resistance in proportion to the energy being detected. By placing the sensor in a simple voltage divider network, the sensor outputs an electrical signal whose amplitude varies in proportion to the energy it senses.

If the robot is required to read the actual energy intensity (analog), an analog-to-digital converter is needed. Analog-to-digital (A/D) converters can measure the electrical signal and output an equivalent binary number output.

A/D convertors will require a microcontroller or digital circuit to function properly and extrapolate the data. In many cases an A/D converter isn't required to read sensor signals. Instead of an A/D convertor, one uses a *comparator*.

As its name implies, a comparator compares two voltages. One is a reference voltage that designers set. The other voltage is derived from the sensor (via the voltage divider). The comparator can output one of two signals: high or low. The high signal is +5 volts and the low signal is 0 volts.

The output signal from the comparator depends on the magnitude of the two voltages on its two input lines. There are three possible choices. The sensor signal is less than the reference voltage, equal to the reference voltage, or greater than the reference voltage.

Comparator example

The best way to learn about comparators is to use one in a circuit. Notice in Fig. 5.1 that the comparator looks much like an op-amp. This is true; comparators are specialized op-amps. The comparator used in our first example is the LM339 quad comparator. This integrated circuit contains four comparators in a 14-pin dip package. Like op-amps, comparators have an inverting and noninverting input. In this particular circuit, the reference voltage is placed on the inverting input (−).

Voltage divider

The voltage divider is a simple but important concept. Using it, you will able to connect most resistance-type sensors to a comparator. The reference voltage is derived from a voltage divider made of two 10K-ohm resistors (see Fig. 5.2A). The Vref in this case will be half of the supply voltage (Vcc), of 5 volts or 2.5V (Table 5.1A). We can make Vref voltage any voltage we require between ground and Vcc by adjusting the two resistance values of the voltage divider (see the following table.

Vcc =5V

To make an adjustable voltage divider, use a potentiometer as shown in Figs. 5.2B and 5.2C. I chose to use the "A"-style Vref because it's simple.

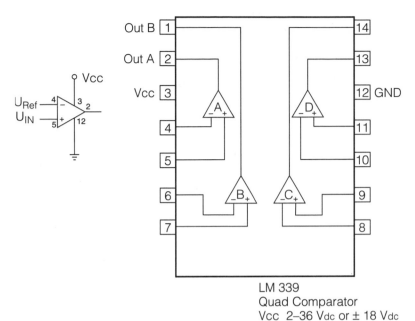

■ **5.1** *Comparator and LM 339 quad comparator IC*

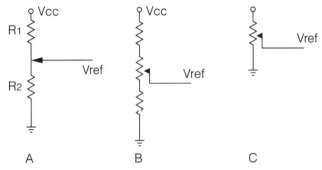

■ **5.2** *Voltage dividers A, B, and C*

Table 5.1A "Two-Resistor Voltage Divider Table"

R1	R2	Vref
1K	10K	4.5 Volts
2.2K	"	4.1 Volts
3.3K	"	3.7 Volts
4.7K	"	3.4 Volts
5.6K	"	3.2 Volts
6.8K	"	2.9 Volts
10K	"	2.5 Volts

The schematic for our test circuit is shown in Fig. 5.3. In place of a sensor, we will use two 1K-ohm resistors and one 5K-ohm potentiometer. By varying the potentiometer, we can adjust the voltage going to the noninverting input (Vin). The output of each comparator is an uncommitted open collector of an NPN transistor. The transistor can sink more than enough current to light an LED, which we use as an indicator. In addition, the output may be used as a simple SPST switch to ground. This feature is useful when needing to trigger a 555 timer used later.

With the circuit wired, let's see what happens. When the input voltage (Vin) is less than the reference voltage (Vref), the output is 0 volts (ground), and the LED is forward based and lit. If we adjust the potentiometer so that the voltage is greater than Vref, the output of the comparator goes high, turning off the LED. You can verify the operation of the comparator by using a voltmeter to measure the voltages at points A (Vref) and B (Vin).

Many people (myself included) feel that this circuit is counter intuitive. I would like the LED to be lit when the sensor voltage is higher than the reference voltage. This can be accomplished by reversing the inputs leads, making the inverting input (–) Vin and the noninverting input connected to Vref. The output function reverses also.

When you don't need too many comparators, you may consider using a CMOS op-amp configured as a comparator. The reason I

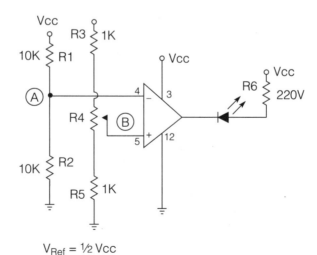

$V_{Ref} = \frac{1}{2} V_{cc}$

■ **5.3** *Comparator test circuit schematic*

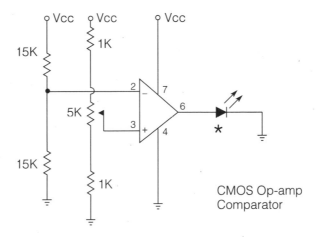

★ Sub miniature LED

■ **5.4** *Comparator op-amp test circuit schematic*

like to use an op-amp is that the op-amp can supply sufficient current to drive an LED or circuit directly (see Fig. 5.4).

Light sensors (sight)

There are a large variety of light sensors: photoresistive, photovoltaic, photodiode, and phototransistor. Light sensors can be used for navigation and tracking. Some robots use an infrared light source and detector to navigate around obstacles and avoid crashing into walls. The infrared source and detector are placed in front of the robot facing in the same direction. When the robot encounters an obstacle or wall, the infrared light is reflected off the surface, causing an increase in infrared detected. The robot's CPU interprets this increased radiation as an obstacle and steers the robot around.

Filters can be placed in front of light sensors to inhibit response to some wavelengths while enhancing response to others. One example of filter use is the flame detectors used in fire-fighting robots. One would try to enhance light from fire while inhibiting light from other sources.

Another example is the use of colored gels to promote color response. One could imagine a robot that separates or picks ripe fruit based on the fruit's skin color.

Photoresistive

CdS (cadmium sulfide) sensors (Fig. 5.5) are photoresistors that can read ambient light. The CdS cell's response to the light spectrum is a close approximation to the human eye (see Fig. 5.6). These are semiconductor sensors without the typical PN junction. The CdS cell displays its greatest resistance in complete darkness. As the light intensity increases, its resistance decreases. Measuring its resistance provides an approximation of ambient light.

Photoresistive light switch

Figure 5.7 is a basic light switch. Because the CdS cell is a resistive-type transducer, it can be placed as a resistor in a voltage divider. When the light intensity increases, the resistance of the CdS cell decreases. This increases the voltage drop on R1 and is seen on pin 2. When the voltage on pin 2 is greater than the voltage on pin 3, the motor turns on. The threshold is adjusted using R1, a 4.7K PC-mounted potentiometer. This is the basic circuit that controls the solar ball project in Chapter 12.

Photoresistive neuron

Figure 5.8 is a light neuron. As the intensity of light increases, the pulse rate becomes faster. The light neuron can provide the clock pulses to a stepper motor controller chip like the UCN5804. As the light increases, the stepper motor turns faster.

Photovoltaic

Solar cells and photodiodes and phototransistors are similar in construction. They all have a light-sensitive PN junction. Solar cells use a wide-area PN junction to produce electrical power in proportion to light intensity.

Photodiodes are usually reversed bias in a circuit. When light strikes the diode's PN junction, it allows current to flow. The photodiode

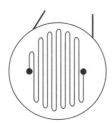

Photo-Resistor(CdS Cell)   ■ **5.5** *Cadmium sulfide cell*

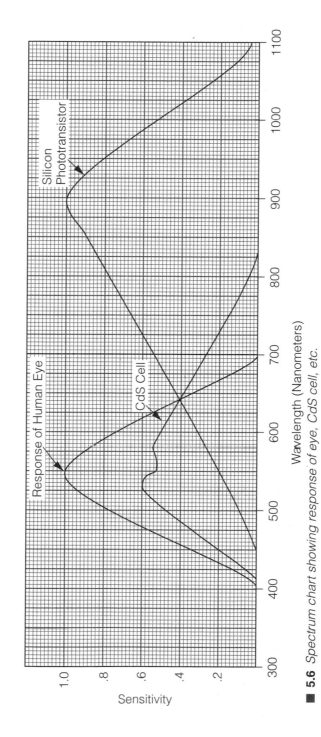

■ **5.6** *Spectrum chart showing response of eye, CdS cell, etc.*

63

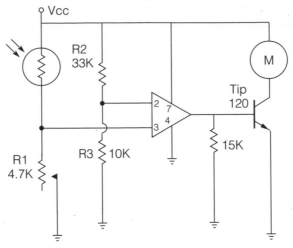

■ **5.7** *Photoresistive light switch*

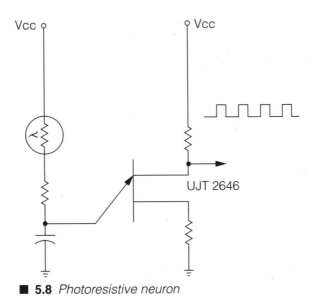

■ **5.8** *Photoresistive neuron*

has a quicker response to the Cds cells and can relay information encoded in the light.

Phototransistors are light-sensitive transistors. The advantage of a phototransistor over a photodiode is that it can provide amplification of the light signal.

Infrared

Infrared (IR) sensors detect low frequency (900-nanometer and longer) light. They deserve special consideration because they are widely used in robotics for tracking, collision avoidance, and communication.

Using Infrared has never been easier. Infrared receiver modules that incorporate modulation detection (shown in Fig. 5.9) are available through a number of electronics distributors. The advantage to these modules is that they only detect IR light oscillating at a specific frequency (usually around 40 kHz).

The 40-kHz waveform can be modulated by another (lower-frequency) signal. The receiver module has also been designed to receive an impressed signal on the 40-kHz carrier wave. This produces a robust communication link. Primarily, the receiver module responds only to the 40-kHz IR signal permitting the receiver to "see" the IR light being transmitted from the transmitter, reject other light sources, then allow the modulation on the 40-kHz wave to be detected.

Infrared collision detector

Figure 5.10 is a drawing of a simple collision detector. As the sensor approaches a solid object, the infrared light reflected back into the receiver increases. The increased infrared light reaches a specific amplitude, where it trips a comparator circuit, informing the robot that there's an obstacle ahead.

■ **5.9** *Infrared receiver module*

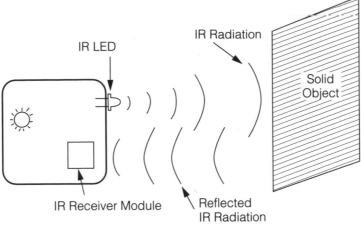

■ 5.10 *Drawing of infrared collision detector*

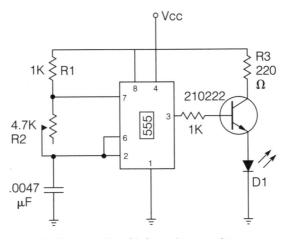

■ 5.11 *Schematic of infrared transmitter*

Infrared transmitter

Figure 5.11 is a schematic of the transmitter. It uses a 555 timer set up in astable mode. Potentiometer R1 is used to adjust the frequency output. The output of the timer (pin 3) is connected to a 2N2222 NPN transistor. An infrared LED is connected to the emitter of the transistor. When you turn on the circuit, don't expect to see any light being emitted from the LED. The infrared light is not detectable by the human eye. Because we are using this in a simple collision detection, there is no need to modulate the 40-kHz signal.

Infrared Receiver

Figure 5.12 is a schematic of the infrared receiver. The receiver module is an Everlite IRM-8420. The center frequency is 37.9 kHz with a bandwidth of 3 kHz (+/– 1.5 kHz). The output is active low. What this means is when the receiver module detects the signal, the output drops to ground. The output is equivalent to an open collector of an NPN transistor (see the insert on Fig. 5.12). The output can sink sufficient current to light an LED. In the test circuit, the LED will light when the module is receiving the signal.

Tuning the transmitter Set up the infrared diode and receiver module next to one another facing in the same direction. The LED must be completely encased in a tube of some sort that only permits the infrared light to leave from the front of the LED. Failure to do this will make using this setup impossible. Note that some plastic materials, while opaque to visible light, are completely transparent to infrared light.

Place a white card about 3 inches in front of the transmitter and receiver. Turn on the circuit. Adjust R1 until the receiver's LED turns on. Then remove the white card completely. The receiver's LED should go off. If it doesn't, the infrared LED on the transmitter may be leaking light from the side and activating the receiver.

Once the unit is working properly, the circuit can be fine-tuned to detect objects at a greater distance. Move the white card back until it just triggers the LED to turn on. Adjust the potentiometer

Receiver Module
1 GND
2 Vcc
3 Output

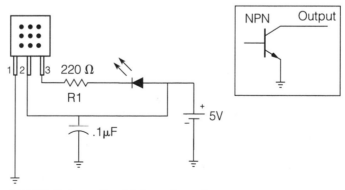

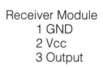

■ **5.12** *Schematic of infrared receiver*

(slightly) on the transmitter so that the LED turns on completely. Keep in mind that it may not be advantageous for the robot to detect objects and/or collisions that are too far away.

DTMF IR communication/remote control system

Other authors have detailed using IR for communication and remote control. Typically the IR transmitter is modulated at a particular frequency and a receiver unit uses a 567 phase-locked loop (PLL) IC. While this works, one must match and tune each transmitter and receiver pair. There is an acceptable work-around to this.

Integrated-circuit chips designed and manufactured for the telecommunications industry are readily available. These inexpensive chips are capable of transmitting and receiving 16 distinct signals, no tuning required. By coupling these chips to standard infrared components, an infrared remote communication/control system can be implemented.

DTMF

DTMF (dual tone multifrequency) was originally developed just over 25 years ago. This was before the U.S. government forced Bell Telephone to break up, allowing the company to expand into other markets. DTMF is commonly known as touch-tone dialing.

The standard DTMF signal is composed of two audio tones generated from a group of eight possible tone frequencies. The eight frequencies are divided into two equal groups, a low frequency group and a high-frequency group (see Table 5-1B). The DTMF signal is an algebraic sum of two tone frequencies, one tone from each frequency group. See Figs. 5.13, 5.14, and 5.15. If we do the math, we see that there are $(4 \times 4 = 16)$ sixteen possible combinations. The low frequencies (R1–R4) are referred to as the row group. The high frequencies (C1–C4) are referred to as the column group.

DTMF encoding

Any combination of frequencies can be obtained using a 4×4 matrix of switches or keypad (see Fig. 5.16). Remember, we are borrowing this technology from the telephone industry. It has been designed for optimum efficiency for less-than-perfect telephone lines.

Standard touch-tone telephones use a 3×4 keypad matrix. This switch matrix provides coding for all of the row frequencies and only three column frequencies (see Fig. 5.17). A 3×4 keypad matrix is more readily available and has been used with all the circuits described.

Not all telephone keypads are made the same, therefore some keypads on the market will not be suitable in these circuits. For instance, some keypads have internal switch wiring that is a little different or includes proprietary ICs. Keep that in mind if the circuit fails to operate properly.

Building a DTMF encoder is simple (see Fig. 5.18). The circuit only requires a keypad, crystal, and 5089 IC. The pinout of the 5089 is shown in Fig. 5.19. Using a standard 3×4 (telephone) keypad will lose the four functional DTMF codes associated with the

■ Table 5-1B Low-Frequency Group

Pin#	Row#	Frequency
R1	Row 0	697 Hz
R2	Row 1	770 Hz
R3	Row 2	852 Hz
R4	Row 3	941 Hz

High-Frequency Group

Pin#	Row#	Frequency
C1	Column 0	1209 Hz
C2	Column 1	1336 Hz
C3	Column 1	477 Hz
C4	Column 3	1633 Hz

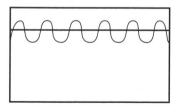

■ **5.13** *Low-frequency tone waveform*

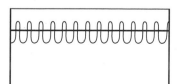

■ **5.14** *High-frequency tone waveform*

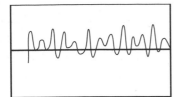

■ **5.15** *Algebraic sum of low and high frequencies (DTMF)*

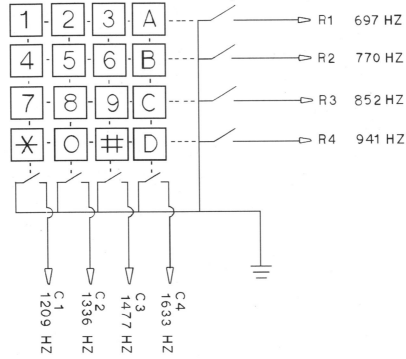

■ **5.16** *4 × 4 keypad matrix showing individual DTMF frequencies*

missing keys, therefore reducing the maximum number of usable channels to 12.

Figure 5.20 is an encoder test circuit using an eight-position dip switch. The dip switch takes the place of the matrix keypad; with it you can test the operation of this encoder circuit and the receiver (decoder) circuit. Notice that when you turn a switch on, you are grounding the pin it is connected to. The pins R1 through R4 and C1 through C4 are active low. Dip switches 1 through 4 are connected to pins R1 through R4, and dip switches 5 through 8 are connected to pins C1 through C4.

The IC can also produce single tones. These are usually generated for testing purposes—for instance, to generate a 1336-Hz tone equivalent to the C2 pin. Ground any two row pins and the C2 pin.

KEYPAD WIRING

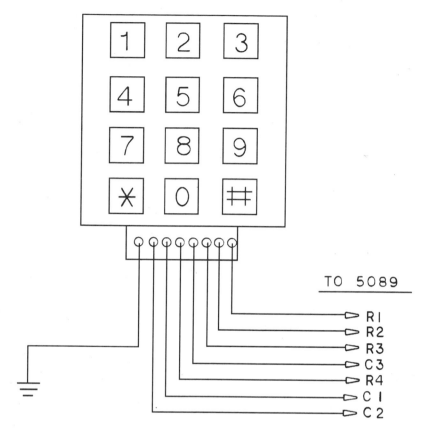

■ 5.17 *Wiring 3 × 4 telephone keypad*

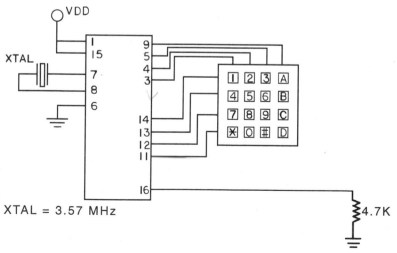

XTAL = 3.57 MHz

■ 5.18 *DTMF encoder circuit using 4 × 4 keypad matrix*

This action will generate a single 1336-Hz signal. The same may be done with the row frequencies. Ground any two column pins with the particular row frequency pin you want to generate.

DTMF decoding

DTMF decoding is just a little more complex than encoding. Again, the simplicity results from the use of a single IC chip, in this case the G8870 (see Fig. 5.21).

The decoding chip has a 4-bit latched output labeled Q1 through Q4. Q4 is the most significant bit (MSB). The current available from the outputs of Q1 through Q4 is sufficient to light a low-current LED. Figure 5.22 is a basic receiving circuit. The output

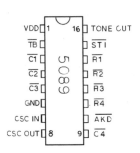

Pin out of 5089 DTMF Transmitter

PIN#	Name	Function
1	VDD	Supply Voltage +5V
2	*TB	Tone disable
3	*C1	Column input
4	*C2	Column input
5	*C3	Column input
6	Vss	Ground
7	OSC1	Clock input
8	OSC2	Clock input
9	*C4	Column input
10	*AKD	Any Key Down output
11	*R4	Row input
12	*R3	Row input
13	*R2	Row input
14	*R1	Row input
15	STI	Single tone inhibit input
16	Tone O/P	DTMF output

* Active Low

■ **5.19** *Pin-out G8870 DTMF encoder IC*

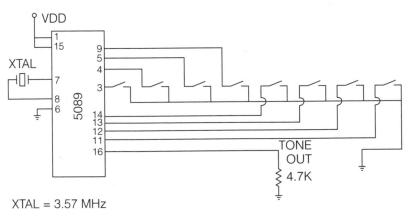

XTAL = 3.57 MHz

■ **5.20** *Schematic encoder circuit using eight-position dip switch*

Pin Out of G8870 DTMF Receiver

PIN#	Name	Function
1	IN+	Noninverting Input
2	IN-	Inverting Input
3	GS	Gain Select
4	Vref	Reference Voltage
5	IC	Internal connection
6	IC	Internal connection
7	OSC1	Clock input
8	OSC2	Clock input
9	Vss	Ground
10	TOE	Tri-state enable output
11	Q1	Output
12	Q2	Output
13	Q3	Output
14	Q4	Output
15	StD	Delay steering output
16	ESt	Early steering output
17	St/GT	Steering input/guard time
18	VDD	Supply Voltage,+5V

■ **5.21** *Pin-out G8870 DTMF decoder IC*

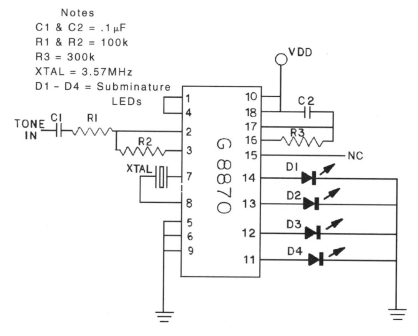

Notes
C1 & C2 = .1 µF
R1 & R2 = 100k
R3 = 300k
XTAL = 3.57MHz
D1 – D4 = Subminature LEDs

■ **5.22** *Schematic receiver circuit with 4-bit binary output*

from Q1 through Q4 lights the LEDs and is a binary number. By looking at Table 5-2, you can determine the binary output that will be displayed on the Q1 through Q4 for all DTMF signals. The way the circuit is wired, the binary "1s" will be represented by lit LEDs.

Table 5-2 DTMF Output Table

| | | Signal Frequency (Hz) | |
Decimal	Binary	Low	High
1	0001	697	1209
2	0010	697	1336
3	0011	697	1477
4	0100	770	1209
5	0101	770	1336
6	0110	770	1477
7	0111	852	1209
8	1000	852	1336
9	1001	852	1477
0	1010	941	1336
*	1011	941	1209
#	1100	941	1477
A	1101	697	1633
B	1110	770	1633
C	1111	852	1633
D	0000	941	1633

Microcontroller

The 4-bit number from the G8870 can be connected directly to input lines of a microcontroller like the Basic Stamp, which can easily read this binary number. (We will get to the Basic Stamp in Chapter 6).

The IR link discussed in the next few paragraphs combined with the Basic Stamp outlined in Chapter 7 will allow users to program communications between mobile robots for games like tag and follow the leader.

Adding a digital display

If reading binary numbers is too cumbersome, we can add a digital numerical display. The output from the chip may also be fed to a BCD to 7-segment decoder chip, such as the 7448.

The 7448 IC is connected to a 7-segment display like the MAN 74 (common cathode). These two chips will provide a digital readout, see Fig. 5.23.

Testing For testing purposes, connect the output from the 5089 chip (pin 16) to the input of the G-8870 chip, using either a keypad or dip switches to generate the DTMF signals. The receiver will display the output via the LEDs or segmented display.

Adding IR transmission

Once the DTMF chips are operating properly, it becomes a simple matter to connect the chips via an infrared radiation (IR) link. The output of the 5089 chip is connected to the base of a common NPN transistor (see Fig. 5.24). A high-power IR LED diode is connected to the emitter of the transistor. Although the IR LED may be connected directly to the output of the 5089, the power output would be small. The NPN transistor allows additional current to power the LED.

Figure 5.25 shows the front end of the IR receiver. An IR phototransistor is coupled to a CMOS op-amp. This combination of components allows the receiver chip (8870) to lock in on the IR radiation from a distance of several feet.

Remote control

Using the IR link, you should be able to press a number on the keypad and see the corresponding number displayed on the digital display. Test the IR link at this point for maximum distance and direction. You should be able to increase distance by placing the IR LED and phototransistor in their own reflectors. The light reflector from an old flashlight will work well.

The remote control begins by adding a 4028 IC. The 4028 is a BCD (binary coded decimal) to decimal decoder, meaning it reads the

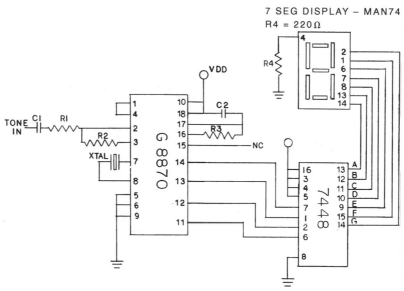

■ **5.23** *Schematic receiver circuit with digital display*

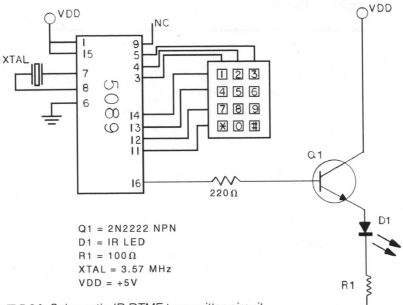

Q1 = 2N2222 NPN
D1 = IR LED
R1 = 100 Ω
XTAL = 3.57 MHz
VDD = +5V

■ **5.24** *Schematic IR DTMF transmitter circuit*

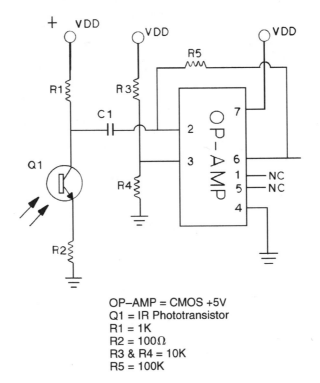

OP–AMP = CMOS +5V
Q1 = IR Phototransistor
R1 = 1K
R2 = 100 Ω
R3 & R4 = 10K
R5 = 100K

■ **5.25** *Schematic front end of IR DTMF receiver*

binary number. (Remember the four LEDs from Fig. 5.22.) The 4028 outputs a single line equal to the decimal equivalent. The 4028 has ten (0 to 9) output lines. Whatever 4-bit binary number is placed on its input lines, it outputs a high signal on that output line (see Fig. 5.26). So, if the binary number "0101," which represents decimal 5, is placed on the inputs, output line 5 will go high on the 4028 IC.

It is not necessary to remove the 7448 and 7-segmented display. The 8870 chip has sufficient output to drive both the 7448 and 4028. The digital display is pretty handy when checking the output from the 4028. For the sake of simplicity, Fig. 5.26 just shows the 4028 connected to the 8870.

The output from the 4028 can be used directly to turn a switch or circuit on or off. However, this isn't an optimum situation. Because as soon as you key another number (channel), the previous channel turns off (brings the line low).

The solution to this problem is a 4013 D-type flip flop (see Fig. 5.27). The flip flop is a basic computer memory datum. In this circuit, it is configured as a divide-by-two counter. Upon receiving the first "on" signal from the 4028, it turns its output line high. When the 4028 brings the line low, which happens when hitting another channel, the 4013 will keep its output line high (latched).

To bring the 4013 output line low, simply key the channel for a second time. The second high signal to the 4013 brings the output

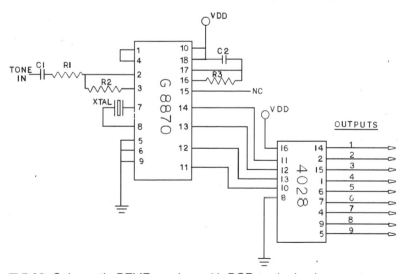

■ **5.26** *Schematic DTMF receiver with BCD-to-decimal converter*

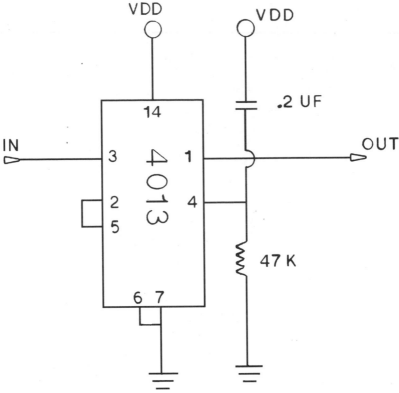

■ 5.27 *Schematic 4013 flip flop*

line low (unlatched). One can continue to bring the 4013 output line high and low by alternately switching the input line high.

Machine vision

To reproduce human vision in a machine is a difficult task. One cannot simply connect a video camera to a computer and expect it to see. Programs (both neural and expert) must capture the video image and process it (extrapolate data). Machine vision has been achieved in limited and targeted areas.

Chapter 1 looked at the Papnet computer, which uses neural software to analyze pap smear slides with a higher accuracy than humanly possible. Other researchers have developed vision systems that can steer a vehicle based on the contours of the road being driven.

Some of the problems that need to be solved to approach human vision (aside from image processing, which is no easy task itself) are stereoscopic mounted video cameras. Some research in this area is taking place at MIT on their humanoid robot, COG. With stereoscopic cameras, two video pictures must be processed, then merged to create a 3D representation. This is the same process used in human 3D vision. To estimate depth, each camera must be mounted on gimbals that allow the cameras to veer in (converge) and focus on an object. The amount of convergence is taken into consideration for judging the object's distance.

Machine vision is a fertile field of development. Currently, most vision systems require a high powered computer dedicated just to vision processing.

Body sense

Body sense provides some information on where one is and what position one is in. Limited body sense can be accomplished in robots by using a variety of tilt switches (see Fig. 5.28). This will at least inform a robot if it is on an incline (decline), flat on its back (stomach), or upside down (right side up). The robot can take appropriate action based on its body sense

Direction-magnetic fields

Using the Earth's magnetic field, an electronic compasses can provide directional information. This will allow a robot to travel in a certain direction or to know which direction it's traveling in.

The simplest sensor in this category is the 1490 digital compass (see Fig. 5.29). The compass is a solid-state hall device.

The digital compass provides four outputs that represent the four cardinal directions: north, east, south, and west. Using a little logic, a total of eight directions can be determined.

The compass is dampened to approximate the speed of a liquid-filled compass. It takes 2.5 seconds for it to respond to a 90-degree displacement. The damping prevents overswinging the direction and prevents switch fluttering when near a switching direction. The device is sensitive to tilting. Any tilt greater than 12 degrees will cause direction errors.

The bottom of the device has 12 leads arranged in four groups of three. Looking at the device from the top, each group of leads is labeled 1, 2, and 3. The leads labeled 1 are connected to Vcc (+5V). Leads numbered 2 are connected to ground. The leads labeled 3 are

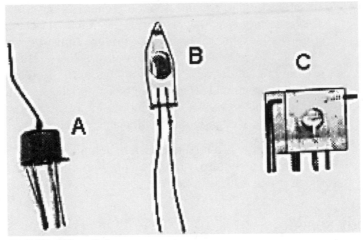

■ **5.28** *Tilt switches*

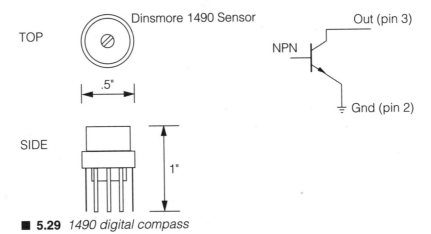

■ **5.29** *1490 digital compass*

the four outputs. The outputs of the digital compass are equivalent to open collectors of an NPN transistor. Being open collects, the outputs are unable to source any current but are capable of sinking enough current (20 mA) to light LEDs.

The test circuit is shown in Fig. 5.30. The sensor will operate with supply voltages ranging from +5 to 18 Vdc. A 9-volt battery is used as a power source and is regulated to +5 volts using a 7805 voltage regulator.

As a rule, try to keep all voltages at 5 volts maximum. This will make it computer safe. For instance, when the digital compass is interfaced to the Basic Stamp computer, if we forget and use a 9-volt power source on the compass, the outputs may damage the I/O inputs.

The test circuit uses four LEDs for display. As the sensor is rotated, each cardinal position on the compass will light one LED. The intermediate directions light two LEDs.

Testing and calibration

To test and calibrate, find north using a standard compass. Rotate the circuit so that one LED is lit. I used the LED furthest away from the sensor for north. If you do the same, the other LEDs will automatically follow the same sequence outlined (see Fig. 5.31). The sequence for my display follows: 1 = on, 0 = off.

LED Sequence

Direction	LEDs	Decimal Equivalent	Inverted
North	0001	1	14
Northeast	0011	3	12
East	0010	2	13
Southeast	0110	6	9
South	0100	4	11
Southwest	1100	12	3
West	1000	8	7
Northwest	1001	9	6

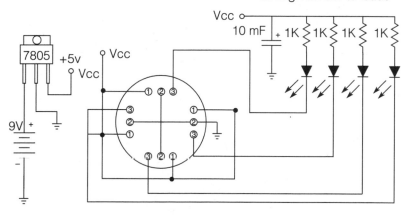

■ **5.30** *Digital compass test circuit using four LEDs*

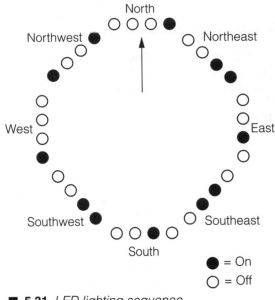

■ **5.31** *LED lighting sequence*

Computer interface

The four output lines from the compass form a 4-bit binary number (nibble) that is easily read by a microcontroller, computer, or electronic circuit. We will hold off on the Basic Stamp circuit until we have introduced the Basic Stamp micro controller in Chapter 6.

1525 electronic analog compass

In most cases, the 1490 directional information is more than sufficient for a robot. However, there will be cases when high-resolution directional information may be important. In this case, one may use the 1525 electronic compass (see Fig. 5.32).

The signal output from the 1525 is much harder to read than the 1490, but the trade-off is that the 1525 electronic compass provides a direction resolution of approximately 1 degree.

The output of this compass is composed of two sine waves, one of which is 90 degrees out of phase with the other (see Fig. 5.32, sine-cosine configuration). The amplitude of each wave correlates to direction. If a 90-degree portion of a sine wave is measured with an 8-bit analog-to-digital (A/D) converter, a compass direction resolution of 1 degree is obtained.

GPS

Using a global positioning system (GPS), a robot can know precisely where on earth it's located. While the need for GPS is not obvious for the amateur robotists, the cost of GPS systems are coming down if the need arises.

Speech recognition

The human ear has an auditory range from 10 Hz to 15,000 Hz. Sound can be picked up easily using a microphone and amplifier, which typically have an auditory range that surpasses human hearing. Sound is a useful tool for robotists.

We use hearing primarily for communication (language), and speech recognition systems are a hot topic in robotics. Because of this, Chapter 7 is devoted to building a speech recognition circuit and interfacing it to a mobile robot.

But don't skip over the following information. Robotic sound systems are pretty useful.

Sound ultrasonics

Sound may be used for games, range finding, and collision and obstacle avoidance. To play a game of robot tag, robots are fitted

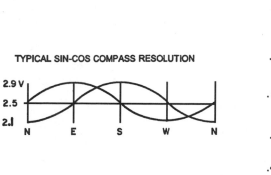

TYPICAL SIN-COS COMPASS RESOLUTION

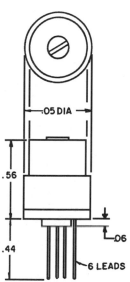

■ **5.32** *1525 electronic analog compass*

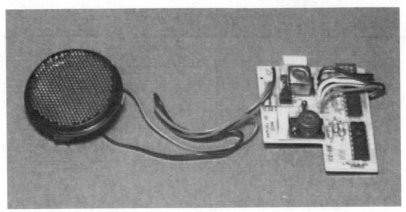

■ **5.33** *Polaroid ultrasonic ranging module*

with a two-tone oscillator and receiver. Each robot can generate and recognize two tones. Let's say the A tone is 3000 Hz and the B tone is 6000 Hz. Tones are generated for one second whenever a robot's bumper switch is activated.

The robot that's "it" generates the B tone whenever it's bumper collides with another robot. A "not it" robot generates the A tone. Upon collision, the "it" robot generates the "B" tone. The "not it" robot, hearing the B tone, changes states and becomes "it." And the "it" robot, hearing the A tone from the "not it" robot, changes states and becomes "not it." Two "not it" robots will both generate the A tone and leave the collision with their states unchanged. Although we are using sound as an example here, be aware that this technique can be applied using infrared light.

Ultrasonics are often used for range finding and collision detection. Many robotists have written on the Polaroid company's ultrasonic modules (see Fig. 5.33). These modules are used in Polaroid cameras to quickly measure the subject's distance from the camera and focus the lens to produce sharp pictures. When interfaced to a microcontroller, the units can accurately measure distance.

If one needs or wants distance measurements for the robot, the Polaroid sensor is the way to go. The ping can measure distance up to 30 feet. The sensor may also be rotated (using a servo or stepper motor), like a radar to build a navigation map and find an obstacle-free path.

Every time the polaroid transducer is energized, there is an audible click from the transducer. I find the constant clicking from the sensor annoying. Although the module is ultrasonic, when the electronics pumps the ultrasonic signal to the transducer, some audible sound is also generated.

To build a basic ultrasonic collision avoidance system is relatively easy, and, being completely ultrasonic, it is silent. The basic operation follows the same scheme used for infrared collision avoidance, except we are using sound instead of light. Figure 5.34 shows the overview schematic. The transmitter transmits a 40-kHz signal to an ultrasonic transducer. Another transducer (receiver) is positioned alongside the transmitter transducer. When the robot approaches a wall or obstacle, the 40-kHz sound is reflected back to the receiver, whose output increases in amplitude. When the output increases beyond the preset point, the comparator trips, relaying that there is an obstacle detected.

Ultrasonic receiver section

The ultrasonic receiver section (see Fig. 5.35) is used to fine-tune the transmitter. The ultrasonic transducers resonant at 40 kHz. If the resonant frequency varies too much (+/–750 Hz) the performance of the transducers degrades rapidly. Fine-tuning the transmitter for optimum resonance frequency is not difficult, provided that you follow the procedure outline. The only piece of equipment needed is a VOM meter capable of reading 2 volts dc.

Because the transducers have a limited bandwidth (resonant at or around 40 kHz), it is unnecessary to add a PLL (phase-locked loop LM567) to the circuit. The transducers naturally reject off-frequency sound.

The receiver section uses a CMOS op-amp. The op-amp is an 8-pin dip that follows the same pin out as the universal 741 op-amp. (Do

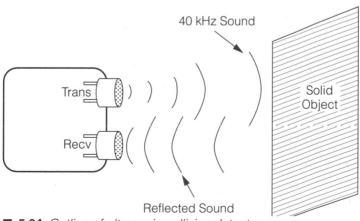

■ **5.34** *Outline of ultrasonic collision detector*

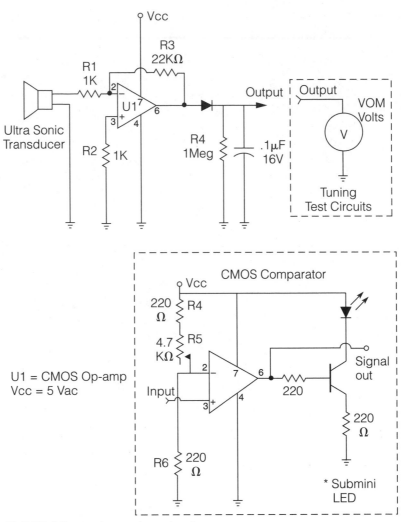

■ **5.35** *Ultrasonic receiver circuit*

not substitute a 741 op-amp). The op-amp is configured as an inverting amplifier with a gain of approximately 22.

Ultrasonic transmitter section

The ultrasonic transmitter (see Fig. 5.36) is built around a CMOS 555 timer set up in astable mode. R2 is a PC-mounted 4.7K potentiometer and is used to adjust the frequency output.

Tuning the transmitter

Set up the ultrasonic transducers so that they are directly facing one another about 4–5 inches apart (see Fig. 5.37). Connect the VOM to the circuit as shown in the insert in Fig. 5.35 (leave the comparator section off). Set the VOM to read DV volts. You will need to read about 2 volts; set the range on the VOM accordingly. Turn on both circuits. Adjust R2 of the transmitter so that you obtain peak voltage output shown on the voltmeter. This should read about 2 volts dc.

Adjusting the CMOS comparator

After tuning the transmitter we need to set the receiver's comparator circuit. Disconnect the VOM from the receiver section and connect the CMOS comparator. Rearrange the transducers so that they are lying side by side about half an inch apart facing in the same direction. Place a flat sided solid object about 3 inches in front of the transducers. Turn on the receiver and transmitter circuits and adjust R5 on the receiver circuit so that the subminiature LED just lights.

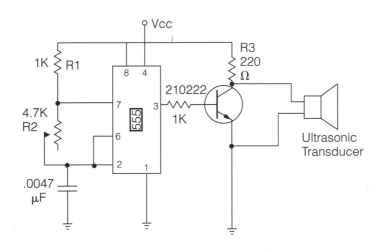

Vcc = 5Volts

* Use CMOS 555 Timer

■ **5.36** *Ultrasonic transmitter circuit*

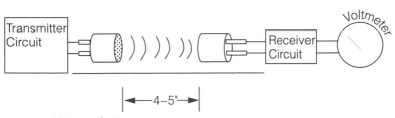

■ **5.37** *Ultrasonic test setup*

To test the circuit, remove the solid object from the front of the transducers, and the LED should turn off. Fine-tune the circuit by placing the solid object 5–6 inches in front of the transducers and readjust R5 until the LED just lights. Note that the receiver is angle sensitive. If the object is held at an acute angle, the ultrasonic sound is reflected away from the receiver. The angles become less crucial the closer the object is to the transducers.

The circuit easily detects solid objects 8 inches away from the transducers. Greater distances can be detected, but, as mentioned earlier, they become angle sensitive. I have the transducers set perpendicularly. You may angle the transducers slightly to obtain different ranging effects.

The circuit provides a TTL high signal that is indicated by the lit LED whenever the circuit detects an obstacle 6 inches away. The TTL signal may be read directly by a neural net or microcontroller.

Arranging the ultrasonic sensors

The obvious use for the ultrasonic system is side (left-right), front, and back obstacle detection. Another use that may not be as obvious is ground detection. Having an ultrasonic sensor facing forward and pointed downward, the sensor reads the ground in front of the robot. If it approaches a cliff or stair, the normally high signal (LED light) goes low, informing the CPU to stop.

Touch and pressure

The human sense of touch has not been remotely approached in robotics. However, there are quite a few simple sensors that can be used to detect touch and pressure. Touch sensors are commonly used for bump and collision sensors.

More sophisticated touch and pressure sensors are used on robotic hands and arms. The sensors allow the robotic hand to grip with enough force to lift without crushing the object.

A simple touch or pressure sensor can be made from electrostatic foam. This is the same foam that integrated circuits (ICs) are packed in to prevent static damage. The foam has a nominal conductivity that changes as the material is compressed. It is important to use low-density (soft) conductive foam because this material is soft and spongy. Figure 5.38 shows a simple touch sensor.

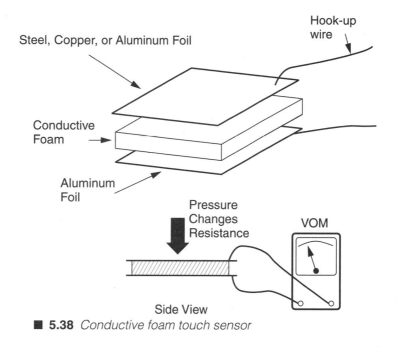

Steel, Copper, or Aluminum Foil

Hook-up wire

Conductive Foam

Aluminum Foil

Pressure Changes Resistance

VOM

Side View

■ **5.38** *Conductive foam touch sensor*

Piezoelectric material

There are a great many piezoelectric sensors. Piezo sensors can detect vibration, impact, and thermal radiation. Pennwall company makes a unique product called piezoelectric film, an aluminized plastic that's been manufactured to render the plastic piezoelectric.

The material is sensitive enough to detect the thermal radiation of a person passing in front of it. Many commercial light sentries sold in hardware stores use piezoelectric film behind a fresnel lens to detect the thermal radiation of a person. These light sentries automatically turn on a light when someone walks into its field of view.

Switches

Momentary contact switches form the foundation of bump sensors, navigation feelers, and limit sensors. There are many types and switch configurations to choose from. Some of the more common switches used in robotics are momentary contact lever and push-button switches (see Fig. 5.39).

■ **5.39** *Momentary contact switches*

■ **5.40** *Bend sensor*

Top View

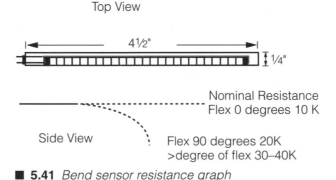

4½"

¼"

Nominal Resistance
Flex 0 degrees 10 K

Side View

Flex 90 degrees 20K
>degree of flex 30–40K

■ **5.41** *Bend sensor resistance graph*

Bend sensors

Bend sensors are passive-resistive devices that increase in resistance as they are bent or flexed (see Figs. 5.40 and 5.41). More commonly used for making virtual reality data gloves to measure the flexing of fingers, these versatile sensors can easily be adapted to robotics. The bend sensor makes an interesting feeler that can inform the robot of an obstacle.

I am reminded of cat whiskers. Cats use their whiskers to determine if a particular passageway is wide enough for the cat to pass through. If the whiskers on both sides of a cat's face touch each side of a passage, the cat will probably not try to pass through it. Bend sensors can be used in a similar manner.

90

Chapter Five

Heat

The most common heat sensor is the thermistor (see Fig. 5.42). This passive device changes resistance in proportion to its temperature. There are positive-coefficient (see Fig. 5.43) and negative-coefficient (see Fig. 5.44) thermistors. As discussed earlier, thermal radiation can also be detected by piezoelectric materials.

Pressure sensor

A simple pressure sensor can be made using conductive foam between two conductors. Conductive foam is used to store static-sensitive integrated circuits (IC). As pressure is applied, the foam squeezes down, changing its resistance between the conductors.

Smell

Currently no sensor exists that can approach the olfactory sense of the human nose. What is available are simple gas sensors that

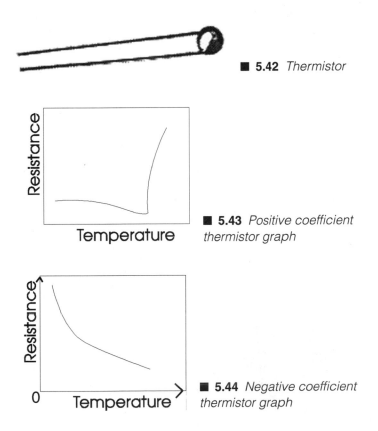

■ **5.42** *Thermistor*

■ **5.43** *Positive coefficient thermistor graph*

■ **5.44** *Negative coefficient thermistor graph*

can detect toxic gases (see Fig. 5.45). The gas sensors can be used to create automatic (robotic) ventilation systems.

A simple sensor setup is shown in Fig. 5.46. The resistive element must be heated to become sensitive, and the sensor incorporates its own heating unit, which is separately powered. The heater requires a regulated +5 volts for proper operation and draws about 130 mA. The resistive element can be read like any other resistive sensor used thus far.

The potential for these gas sensors is greater than what might be implied in the simple schematic. The gas sensors are not precise instruments; their response varies slightly from device to device. This "analog" property can be used to create a more sensitive smell detector.

Let's arrange eight sensors. The resistive element from each sensor is connected to an A/D convertor. A comparator circuit wouldn't do in this situation because precise and subtle variations in response are what we are looking for. To calibrate the device, a small amount of a known gas (smell) is released by the eight sensors. The response of each detector is measured by the A/D convertor and recorded by the main computer. Since the response of the detectors will vary, an eight-number pattern is created for each smell.

Pattern matching is well established in neural networks. A neural network can not only measure scents, but it can also recognize different smells.

Humidity

Passive resistive humidity sensors are a relatively new product that can be purchased.

Testing sensors

When designing and building sensor systems, it's a good idea to test them before committing the system on a robot. One method

■ **5.45** *Toxic gas sensor*

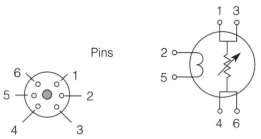

Pins

Bottom View Sensor

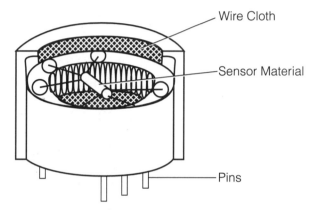

Electrical Schematic Sensor

Wire Cloth

Sensor Material

Pins

Cutaway View — Gas Sensor

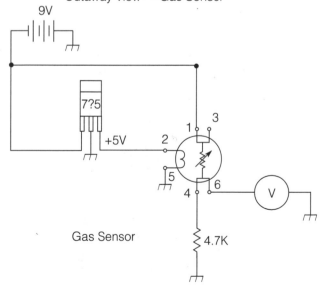

Gas Sensor

■ **5.46** *Toxic gas sensor test circuit*

that I used was to build a small, mobile robot whose only function is to test sensors. That way, reliability and response time can be determined before committing the sensors on a more elaborate robot.

The robot can test bump switches, light switches, bend sensors, and infrared and ultrasonic obstacle avoidance sensors. Other types of sensors may require a different test bed.

Tester

Tester is the name I've given this small robot. The foundation of the robot is a small electric car that can be purchased for less than $10.00 (see Fig. 5.47).

The Tester schematic is shown in Fig. 5.48. The sensor connects to the trigger input of a 555 timer set up as a monostable pulse generator. The output (pin 3) of the 555 remains low until a negative pulse on the number 2 pin triggers its operation. Once triggered, the output (pin 3) of the 555 goes high for approximately one second.

The output of the 555 connects to a 2N2222 NPN transistor. An output is taken off the emitter of the transistor and connected to

■ **5.47** *Tester*

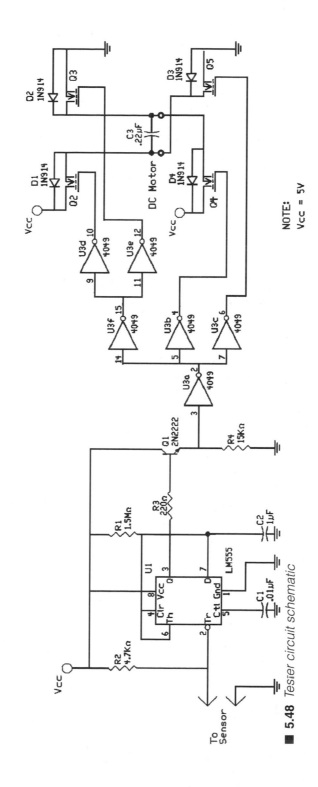

5.48 *Tester circuit schematic*

NOTE:
Vcc = 5V

a buffer on the 4049 hex inverting buffer IC. The buffers on the 4049 chip are connected to a four-MOSFET transistor H-bridge that controls the drive motor.

When the output of the 555 timer is low, the H-bridge powers the robot's drive motor forward. The sensor to be tested is connected to the trigger input, pin 2 on the 555 timer. The sensor is wired in such a way as to cause a negative pulse (goes to ground) when it's activated or tripped. The negative pulse on pin 2 causes the output of the timer to go high for one second, which reverses the motor direction for one second. Tester can be used to check a variety of sensors and transducers.

Improving the Tester robot

When I designed Tester, I had imagined that most of the sensors I would test and use would be tiny miniature modules. This was not the case. In the process of prototyping different circuits, I rarely had the time to produce a PCB, let alone miniaturize the circuit.

If I were to build another Tester robot, I would use a much larger electric car as a foundation. Having a lot of room to work on the robot makes it easier to secure different types of sensors and circuits.

Parts for the projects outlined in this chapter are available from:

Images Company
POB1 40742
Staten Island, NY 10314
(718) 698-8305

Intelligence

6

INTELLIGENCE PACKAGED IN A ROBOT TAKES ONE OF TWO forms: rule-based (expert) or neural. It's possible for both forms of intelligence to work in tandem. This synthesis of intelligence will be commonly used in robotics to create a robust intelligence system.

Expert (rule-based) intelligence systems are familiar to most people; these are programs written in high-level or low-level languages like C++, BASIC, and assembly. Neural systems, on the other hand, use electronic neurons and feedback to control the robot. The subsumption architecture (neural) pioneered by Rodney Brooks at MIT is discussed in Chapter 17.

This chapter focuses on rule-based programs (expert) systems and microcontrollers. Keep in mind that it's possible to mimic neural systems using expert-systems programming. Currently, 95% of all neural network software on the market runs on existing rule-based computers, using rule-based programming.

BASIC Stamp microcomputer

Adding intelligence in the form of a computer to a small robot or robotic system has never been easier. There are numerous single-board computers (SBCs, also known as microcomputers) available that can do the job. The advantages of using a microcomputer are numerous: decision making based on directly reading switch and sensor status, powering dc motor drives (using dc current or PWM), controlling servo motor positioning, stepper motor, etc. Naturally, one can program the robot (microcontroller) to respond to the sensor readings. This makes prototyping robotic behaviors easier.

A single-board computer (SBC) is exactly what its name implies—an entire computer on a single printed circuit board. The main component is called a *microcontroller*. SCBs are used in diverse applications, from antilock braking systems in cars to washing-machine controllers. Typically, microcontrollers are programmed in assembly language. Assembly is a difficult low-level (high-speed) language.

The development packages for microcontrollers can be complex. Development packages usually include an editor to write the program, an assembler that transforms the program into machine code that can be downloaded into a chip, and finally a development board that connects the chip to the host computer.

After the chip is programmed, it is removed from the development board and placed in its application circuit board. As you can imagine, the total cost for these development packages can run into many hundreds of dollars.

The solution to these obstacles can be found in a low-cost SBC that runs BASIC programs. The cost for a complete development system is $114, which includes one SBC. The name of the SBC is the BASIC Stamp. The name refers to the small size of the device. The complete, ready-to-run program costs $34.

BS1 or BS2

Stamps come in two flavors. The BASIC Stamp 1 (BS1) is the original product. It has a small 256-byte working memory and 8 I/O lines. The BS2 is a later product. The advantages of the BS2 are twice as many I/O lines (16 I/O lines) and a working memory of 2K. Obviously, one has a lot more room to expand with the BS2 system. We will work with both systems. Much of what is learned working with the BS1 can be transferred over when working with the BS2 system.

BS1

The BASIC language used in the Stamp is derivative of standard BASIC. It has a few quirks, but it is familiar enough for most users to begin writing programs right out of the box. In this chapter, we examine the Stamp's hardware and software, and we use this SBC to create a few innovative controller projects.

About the BASIC Stamp

The heart of the BASIC Stamp is a PIC 16C56 microcontroller (see Fig. 6.1). The support components include a 256-byte EEPROM, 4-MHz resonator, 5V power supply, and a few resistors. Figure 6.2 is a drawing of the Stamp, which is constructed on a 1" × 2" double-sided PC board. A prototyping area of 36 plated through holes is provided for the experimenter.

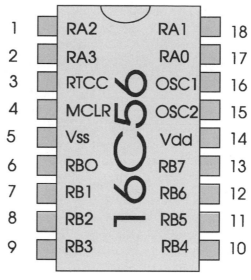

1	RA2	RA1	18
2	RA3	RA0	17
3	RTCC	OSC1	16
4	MCLR	OSC2	15
5	Vss	Vdd	14
6	RB0	RB7	13
7	RB1	RB6	12
8	RB2	RB5	11
9	RB3	RB4	10

■ **6.1** *PIN 16C56 microcontroller*

Soldered 6" wires to the eight I/O lines plus the +5V power and ground leads to the BASIC Stamp's prototyping area.

We then fed the wires to a solderless breadboard. This makes experimenting with the Stamp a snap.

The Stamp is powered by a 9V battery that connects directly to the circuit board. Power is regulated down to +5V by the onboard 7805 voltage regulator.

Eight I/O lines

The microcontroller comes programmed with a BASIC language interpreter. It has eight programmable I/O (input/output) lines, and the lines may be programmed to function as dedicated input or output lines. In addition, it is possible to change an I/O line's status (either input or output) on the fly from the running BASIC program.

The Stamp's I/O pins can each source 20 mA or sink 25 mA. It is recommended that the total sink current into the I/O lines not exceed 50 mA, while the total source current delivered by the I/O lines should be limited to a maximum of 40 mA.

The development package includes software (editor and assembler), programming cable, manual, application notes, and one BASIC Stamp computer (see Fig. 6.3).

The development package requires a minimal IBM PC or compatible computer to run, and the computer needs to equipped with at least one disk drive to load and run the Stamp program. A parallel (printer) port is needed for the programming cable, which connects the computer to the stamp. The computer should be running MS-DOS 2.0 or greater and have at least 128K of free RAM. The programming cable has a DB-25 connector on one end that plugs into a 3-prong header on the Stamp computer.

Using the BASIC Stamp

To use the BASIC Stamp, you simply connect it to the computer, apply power, and run the Stamp's software. The editor is used for writing programs. The four menu items on the top of the editor's screen are activated by pressing the ALT key plus the first letter of the menu item. The four menu items are: RUN, LOAD, SAVE, QUIT. The editor is simple ASCII text-writing software. It saves the Stamp programs as ASCII text files. You may use a more powerful word-processing software, as long as it's capable of saving its files as unadorned ASCII text. The ASCII text files may be loaded into the Stamp's editor and downloaded into the Stamp.

To load a program, you press the ALT and L keys simultaneously. A window opens for you to enter the name of the program you

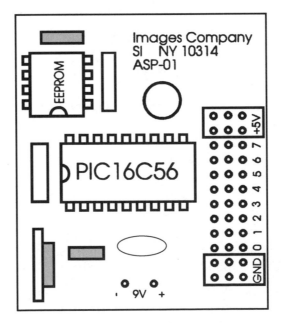

■ **6.2** *BASIC Stamp development package*

■ **6.3** BASIC Stamp.

want to load. To download your program into the Stamp, press the ALT key and R keys simultaneously.

When RUN is initiated, the software first checks the BASIC program and reports any errors it may have found. Then it looks to see if the hardware (stamp) is connected. If everything is okay, it begins downloading the program into the Stamp. A bar-graph appears on the screen that shows the progress of the download and gives an approximate estimate of the remaining memory available.

After the program is completely downloaded, the Stamp reboots itself and begins executing the program immediately. In fact, whenever power is applied to the Stamp, it boots the program and begins execution. The Stamp's serial EEPROM is used for program storage. The EEPROM retains the downloaded program even after power is removed from the circuit, and it can store a program without power almost indefinitely (20 years). Fortunately, you're not stuck with the first program you load into the EEPROM because the EEPROM is re-programmable. So you can make any changes you wish to the program, then download the program into the Stamp again. The Stamp will accept and begin executing the new program immediately.

The EEPROM can be reprogrammed an estimated 1,000,000 times before it wears out, and stopping and starting the microcontroller doesn't place any wear on the EEPROM chip. Wear accumulates only by reprogramming the chip and using the WRITE command in BASIC.

The BASIC language

The development kit's manual illustrates and explains each BASIC command that can be used in the Stamp. A brief summary of the BASIC commands is given in Table 6.1 at the end of this chapter to help you follow the sample application programs presented.

Nitty gritty

In previous chapters I've mentioned the microcontroller often and stated how easy it will be to apply to particular tasks. The following is a list of robot applications. So let's get started.

Reading switches

Switch status can be read using the command INPUT pin. The word pin is a variable 0–7 that corresponds to one of the 8 I/O lines on the Stamp. The command may also be used for reading the output of a comparator. Switches must be configured on the robot to give either a high or low signal. Either signal may be read by the microcontroller and acted upon.

Reading high A normally open momentary contact switch can be configured to give a high signal on closing, as shown in Fig. 6.4. The following program will read the switch status and light a subminiature LED when the switch is closed.

```
REM Test Switch High
input 7      'Make Pin 7 input
output 0     'Make Pin 0 output
start:
if pin7 = 1 then LED
low 0
goto start
LED:
high 0
goto start
end
```

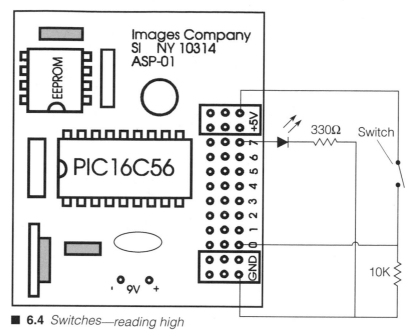

■ **6.4** *Switches—reading high*

While this program may appear trivial, it conveys useful basic instructions. If one replaces the LED with an NPN transistor that controls a dc motor, the high signal could switch on and off a dc drive motor.

Reading low Figure 6.5 shows a switch that produces a low signal when closed. The program is almost identical to the first except for line 4, "if pin0 = 0 then LED."

```
REM Test Switch Low
input 7     'Make Pin 7 input
output 0    'Make Pin 0 output
start:
if pin7 = 0 then LED
low 0
goto start
LED:
high 0
goto start
end
```

Reading comparators

Comparators were covered in Chapter 5. The information on comparators will not be repeated here. If you are not clear on how the comparator examples work, review the comparator section in Chapter 5.

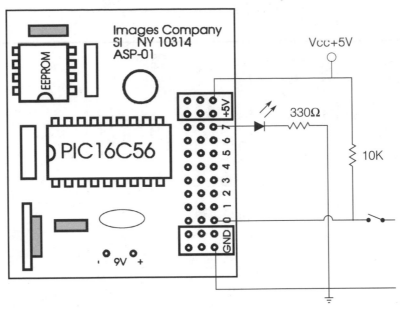

■ **6.5** *Switches—reading low*

The output of comparators may be read the same way as switches. Since the output of a comparator (LM339) is equivalent to an open collector of an NPN transistor, it is usually brought high by using an external pull-up resistor to Vcc (see Fig. 6.6). The comparator is read from the Stamp using the same program that reads a low signal from a switch.

If a single-ended supply op-amp is configured as a comparator, see Fig. 6.7. Its output may be low (0 volts) or high (+5 volts) and may be read using either switch program.

Reading resistive sensors

The BASIC Stamp is able to read resistive sensors (5–50K) directly. The types of resistive sensors available are: photoresistors (CdS cells), thermistors (PTC and NTC), toxic gas sensors, bend sensors, humidity sensors, etc. The limitation is that the resistance value of the sensor must lie between 5K to 50K.

The command to read a resistive sensor is:

```
POT pin,scale,variable
```

☐ Pin is the variable/constant (0–7) that specifies the I/O pin used.

☐ Scale is a variable/constant (0–255) that reduces the 16-bit result of the POT command to an 8-bit value. An 8-bit value is the maximum value that can be held in a 1-byte variable. The result is multiplied by the fraction (scale/256). A scale value of 128 would reduce the result by 50%.

☐ Variable is used to store the final result of the POT command.

Finding the best scale value The Stamp has a run-time procedure for finding the best scale value for the sensor. Press ALT-P keys while in the Stamp's editor. This will open a window; select the I/O pin the sensor is connected to. The editor will download a small program into the stamp. Next a window will appear with two numbers: scale and value. Adjust the sensor so that the smallest number is shown for scale. Use this scale number in the POT command line.

To test the scale number, there is an optional step you may take if you want. Press the spacebar; this will lock the scale and allow the stamp to continually read the sensor. Sensor values can vary between 0 and 255 max (8-bit number).

Photoresistive sensor

Figure 6.8 is the schematic for the photoresistive sensor program. The Stamp determines the resistive value by timing how long it takes to charge the capacitor.

```
Rem Photo-Resistor Test Program
start:
output 1     ' Make Pin 1 an outputoutput 2     ' Make Pin 2 an
output
pot 0,43,b2 ' Read resistor on Pin 0
```

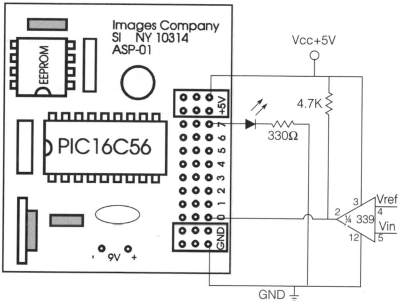

■ **6.6** *Reading low signal from LM339 comparator*

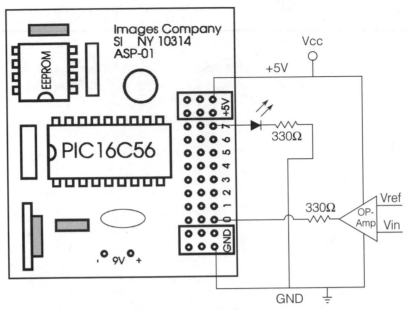

■ **6.7** *Reading high or low signal from op-amp comparator*

```
if b2 > 100 then 11    ' If more than 100 lite LED 1
if b2 < 100 then 12    ' If less than 100 lite LED 2
11:     ' Lite LED 1 routine
pin1 = 1:pin2 = 0
goto start  ' Do Again
12:     ' Lite LED 2 routine
pin2 = 1:pin1 = 0
goto start  ' Do Again
```

Multiple resistive sensors

The BASIC Stamp can read resistive values off multiple pins. In this second photoresistive sensor program, the values from two photoresistive sensors are derived. The program can be modified to track a light source or act as a steering control on a mobile robot to follow a light source. Figure 6.9 is the schematic for the test program.

```
Rem Dual Photo-Resistor Test Program
Rem May be used to steer robot toward light source.
output 2    ' Make Pin 2 output
output 3    ' Make Pin 3 output
start: ' Start
pot 0,100,b4' Read resistor on Pin 0
pot 1,100,b5' Read resistor on Pin 1
let w5 = b4 + 24 ' Create reading range
let w6 = b5 + 24 ' +/- 24
if w5 < b5 then 11    ' Check w5 if out of range
if w6 < b4 then 12    ' Check w6 if out of range
pin3 = 1: pin2 = 1    ' If in range lite both LED's
goto start  ' Do again
11:     ' Routine to lite LED
pin3 = 1:pin2 = 0' on pin 3
goto start  ' Do Again
```

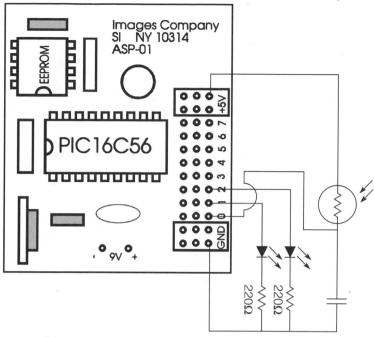

■ **6.8** *Schematic for photoresistor sensor program*

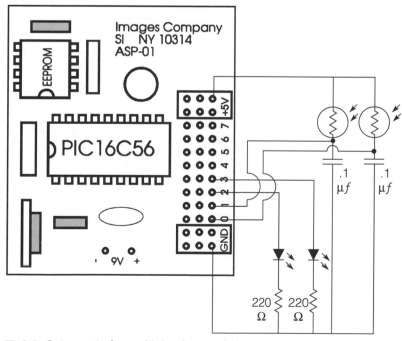

■ **6.9** *Schematic for multiple photoresistive sensors*

```
12:    ' Routine to lite LED
pin2 = 1:pin3 = 0' on pin 2
goto start   ' Do again
```

Tracking a light source

To track a light source, the photoresistors are mounted on the shaft of a gear box motor. The output signals from the stamp control the gear box motor using an H-bridge. (See Fig. 6.10. Also see H-Bridge motor control further on in this chapter.) Adjust the logic in the program so that when both photoresistors are evenly illuminated, no power is provided to the motor.

Light-following robot

The light-following robot uses two miniature gear box motors. The gear box motors provide steering and forward motion. When both motors are switched on, the robot travels forward, and steering is accomplished by switching one motor off while the other motor remains on (see Fig. 6.11).

Controlling servo motors

Controlling servo motors is easy. The following program shows how simply it can be accomplished. Two switches connected on

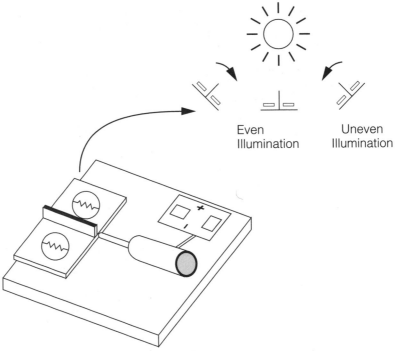

Even
Illumination

Uneven
Illumination

■ **6.10** *Light tracking using gearbox motor*

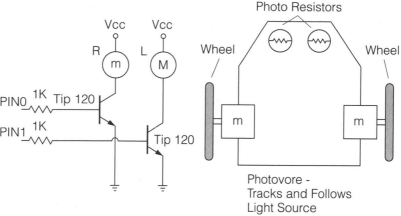

■ 6.11 *Light tracking mobile robot*

pins 4 and 5 control the direction the servo motor will turn (see Fig. 6.12). If neither switch is pressed, the servo motor is placed in its centered position.

Servo motors are controlled using 1–2 millisecond pulses sent to the motor 50–60 times a second (see servo motors in Chapter 4). The Stamp can output pulses in 10-microsecond units.

Program description: Dirs = %00001111 sets pins 0 to 3 as output and pins 4–7 as input pins.

Pulsout 0,150 sends an output pulse of 1.5 ms on pin 0.

Pause 20 causes the program to wait 20 ms before continuing. This delays the recurring pulse signal so that approximately 50 (Hz) pulse signals are delivered to the servo every second.

The pulse width is produced on pin 0 (see Fig. 6.13). The Stamp can easily control two or more servos. A similar program controls the walker robot in Chapter 11.

Test servo motor program

```
dirs = %00001111
loop:
pulsout 0,150
pause 20
l1:
if pin4 = 1 then l2
if pin5 = 1 then l3
goto loop
l2:
pulsout 0,75:
pause 20 : goto l1
l3:
pulsout 0,225:
pause 20 : goto l1
```

Servo sweep

After you've gone through the trouble of building a sensor system, I'm sure you want to use it to its fullest capacity. For instance, with the ultrasonic collision detector, one could build two or three sensor systems and mount them on the front of a mobile robot. This will blanket coverage for the entire front of the robot.

Another method is to mount the a single sensor on a servo motor positioned on the front of the robot. The servo motor can then sweep the sensor in a 120–180-degree circumference in front of the robot. The robot's speed cannot be so great that it travels past the sensor collision detection range in the time it takes for the servo/sensor to return.

When the sensor detects an object, the object position is calculated from the servo pulse signal. This information is used to determine the direction the robot should take to avoid the collision.

Fuzzy logic

A pseudo fuzzy logic subroutine determines the direction the robot should take to avoid collision. Fuzzy logic differs from logic in that the boundaries between sets is fuzzy. This is best explained with an example.

Let's divide a random group of people into three subgroups that are determined by height. The first group we will call "short people," the second group "medium People," and the third group "tall people." For the sake of the example, let's state that the median height for the short people group is 5 feet. Median height for medium group is 5 feet 6 inches, and the median height for the tall people group is 6 feet.

If we plotted the distribution curve for each group of people, the curve would resemble Fig. 6.14. To simplify our discussion, we will replot the distribution curve to a similar-sized triangle (see Fig. 6.15).

Figure 6.16 graphs the three height groups of people. The shaded areas of the triangles represent where the groups overlap. A person whose height lies in the shaded areas can be the tallest person in one group or the shortest in the other. What group that person belongs to becomes a matter of interpretation.

Standard logic abhors shaded areas. It either is or is not part of a group. Because of this, fuzzy logic was created to interpret shaded areas between groups. In true fuzzy logic circuits, a chip interprets the areas between groups and provides an output.

I've programmed a simple pseudo fuzzy logic routine into the basic sweep program. The purpose is to determine which way and how much the robot should turn to avoid an object. The graph representing the fuzzy logic is illustrated in Fig. 6.17. The numbers on top of the graph (75, 150, and 225) represent in microseconds the position (rotation) of the servo motor at the point it detects an obstacle. At the bottom of the graph, the decision, based on position, is provided.

Figure 6.18 shows the unltrasonic detector module secured to the shaft of a servo motor for testing. The schematic for the circuit is shown in Fig. 6.19.

Servo sweep program

```
dirs = %00011111
'initialize
```

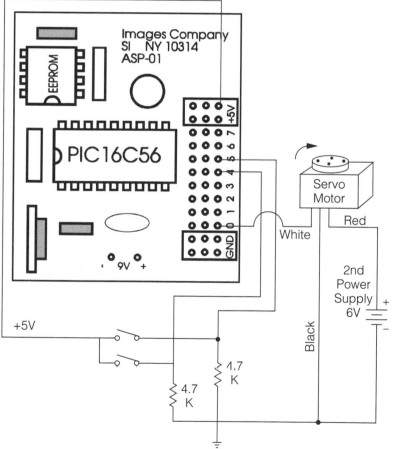

■ **6.12** *Schematic for controlling servo motor*

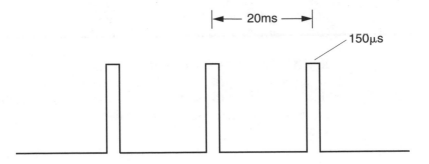

Servo pulse signal from PIN Ø

■ **6.13** *Pulse width for servo motors*

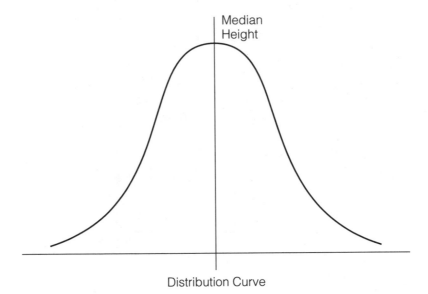

Distribution Curve

■ **6.14** *Distribution curve for people in one height group*

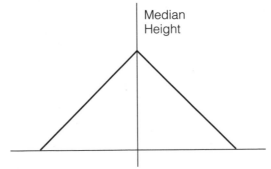

■ **6.15** *Replot of distribution curve into triangle*

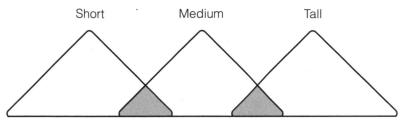

75 – 100   Turn Right
101 – 115   Turn Right Sharply
116 – 185   Reverse and Turn Sharply
186 – 200   Turn Left Sharply
201 – 225   Turn Left

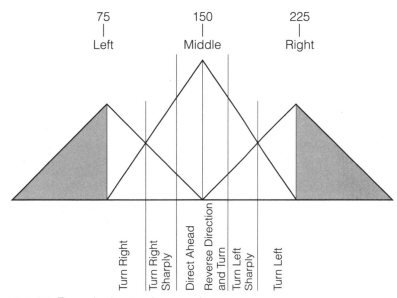

■ **6.16** *Graphs the three height groups showing overlap*

■ **6.17** *Fuzzy logic steering graph*

```
b3 = 0'use b3 as flag to remember
b1 = 75   'sweep direction
sweep:
pin4 = 1: pin2 = 0  ' move forward
pulsout 0,b1    ' servo signal
pause 20  ' transmit signal at 50-60 Hz
b3 = 1' flag
if pin5 = 1 then collision' object detected?
b1 = b1 + 1' increment servo signal
```

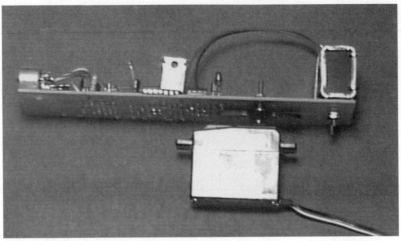

■ **6.18** *Picture of ultrasonic detection module*

```
if b1 => 225 then sweepback     ' end of sweep?
goto sweep ' if not do again
sweepback: ' start sweep back
b1 = b1 - 1' decrement servo signal
pulsout 0,b1     ' servo signal
pause 20    ' transmit signal 50-60 Hz
b3 = 2' flag
if pin5 = 1 then collision' object detected?
if b1 <= 75 then sweep     ' end of sweepback?
goto sweepback  ' if not do again
collision: ' detected collision
```

Rem Pseudo FUZZY LOGIC (see text).

```
if b1 < 101 then r     'turn right
if b1 < 116 then rs  'turn right sharply
if b1 < 186 then rts 'reverse and turn sharply
if b1 < 201 then ls  'turn left sharply
b2 = 3'turn left
goto turn
ls:     ' turn left sharply
b2 = 3
goto turns
r:     ' turn right
b2 = 1
goto turn
rs:     ' turn right sharply
b2 = 1
goto turns
rts:   ' reverse direction & turn
pin4 = 0:pin2 = 1     ' reverse direction
b2 = 3' turn
goto turns
turn:
high b2
pause 500
low b2
if b3 = 1 then sweep ' return to sweep
if b3 = 2 then sweepback  ' return to sweepback
```

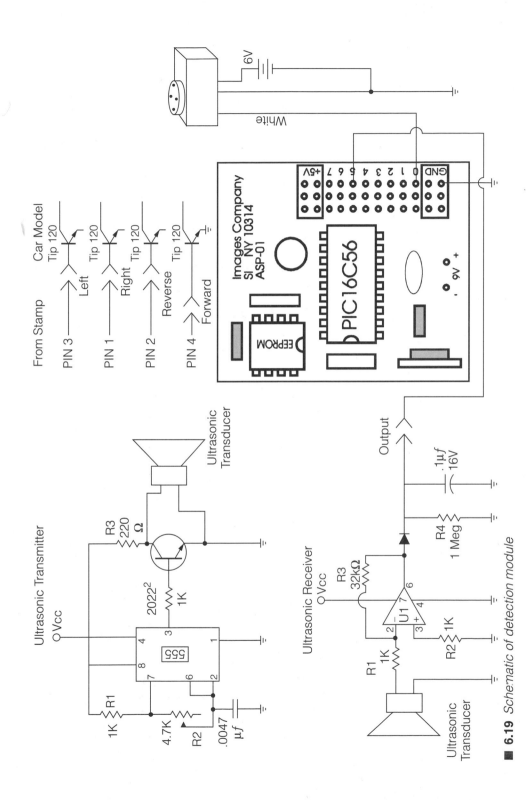

6.19 *Schematic of detection module*

```
turns:
high b2
pause 1000
low b2
if b3 = 1 then sweep ' return to sweep
if b3 = 2 then sweepback  ' return to sweep back
```

Platform interface

The platform I have used for the servo sweep is an old radio-controlled (RC) electric car (see Fig. 6.20). All the RC equipment was stripped from the car, leaving the wires that control the motor and steering. This particular model can be controlled using four transistors. Each transistor controls one function: forward, reverse, turn left, turn right. Consequently, I wrote the sweep program to activate each car function accordingly. With a little work, it should not present too much of a problem to interface this circuitry to most other platforms.

Other uses for the sweep program

Electronic bloodhound An electronic bloodhound robot can be created using the toxic gas sensor. The toxic gas sensor/servo is positioned in front of the robot and close to the ground. When the servo sweeps, it sweeps the gas sensor through a 100-degree arc in front of the robot.

To test the robot, one places an ammonia trail for it to follow. Wet a piece of cotton with ammonia and write a trail on a smooth-surface floor. Place the robot at the beginning of the trail and let it run.

The logic within the sweep program must be reversed. Instead of collision avoidance, it must follow and track the scent. The infrared collision detector can be swapped for the ultrasonic collision detector without too much trouble.

Controlling dc motors

We established in Chapter 4 that an H-bridge is an ideal method of controlling dc motors. In addition to turning on and reversing motor direction, the microcontroller can be used to control the H-bridge (see Fig. 6.21). The Stamp can also provide pulse width modulation (PWM) to control the motor's speed.

While the Stamp has a PWM command in its syntax, it is not a command that I would use. The reason is that the command is limited to a maximum of 255 PWM cycles. While a work-around is possible, it's just as easy to write PWM code.

Dc motor control

```
dirs %00001111
start:
```

■ **6.20** *Electric car platform for servo sweep*

```
if pin4 = 1 then forward
if pin5 = 1 then reverse
stop:
pin0 = 0
pin1 = 0
goto start
forward:
pin1 = 0
pin0 = 1
pause 1
goto start
reverse:
pin0 = 0
pin1 = 1
pause 1
goto start
end
```

Dc motor control using PWM and H-Bridge

```
dirs = %00001111
REM 3 speed; slow, medium and full
speed.
if pin4 = 1 then full
if pin5 = 1 then medium
if pin6 = 1 then slow
stop:pin0 = 0
goto speed
full:
pin0 = 1
pause 1
goto speed
medium:
```

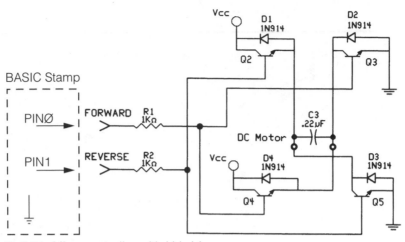

■ **6.21** *Microcontroller with H-bridge*

```
pulsout 0, 250
pause 1
goto speed
slow:
pulsout 0, 120
pause 1
goto speed
```

With a little more logic, the PWM program could incorporate directional control using an H-bridge.

Controlling stepper motors

The BASIC Stamp is capable of directly controlling a stepper motor. To accomplish this, four of the Stamp's output pins are connected to each stepper motor winding through a transistor. The output pins are turned on and off using the stepper motor sequence.

Another way is to use a dedicated stepper motor controller chip such as the UCN5804 and have the Basic Stamp control the stepper motor through the chip. This is the approach we shall use (see Fig. 6.22). This approach saves us from writing full-step and half-step forward and reverse sequences.

The speed of the stepper motor can be controlled by varying the pause statement. Smaller pause values will make the motor rotate faster. Currently the stamp sends out a 50-Hz square wave that half-steps a 1.8-degree motor at 7.5 rpm.

Stepper motor controller

```
dirs %00001111
start:
high 0 'Bring pin 0 high
pause 10    'for 10 ms
```

```
low 0   'Bring pin 0 low
pause 10   'for 10 ms
if pin4 = 1 then reverse   'check direction
if pin4 = 0 then forward   'check direction
reverse:   'reverse direction
pin1 = 1
goto start  ' do again
forward:   ' forward direction
pin1 = 0
goto start  ' do again
```

More applications

The development package includes a set of application notes that contain schematics and programs for a diverse group of projects. The projects include:

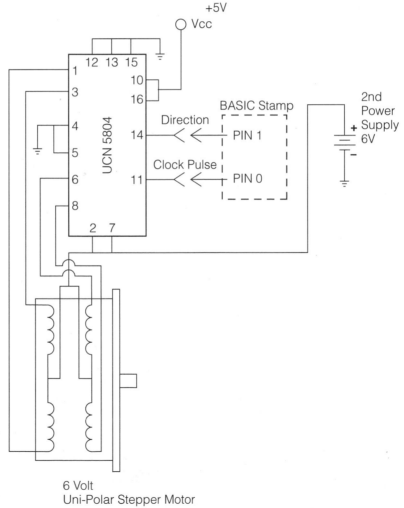

6 Volt
Uni-Polar Stepper Motor

■ **6.22** *Schematic basic stamp stepper motor controller*

☐ LCD user-interface terminal

☐ Interfacing an A/D converter

☐ Hardware solutions for keypads

☐ Controlling and testing servos

☐ Practical pulse measurements

☐ A serial stepper controller

Basic commands

■ **Table 6-1**

Basic command	Explanation
BRANCH	Branch to an address specified.
BUTTON	Debounce switch on I/O line.
DEBUG	Used to "print" variables and status of I/O lines to PC.
EEPROM	Stores values in EEPROM before loading, BASIC program.
END	Enter sleep mode, low power consumption.
FOR NEXT	Loop.
GOSUB	Jump to subroutine.
GOTO	Jump to address specified.
HIGH	Make output of pin specified high (= 5V).
IF THEN	Compare variable content and branch.
INPUT	Make specified pin an input.
Basic command	**Explanation**
LET	Assign a value to a variable.
LOOKDOWN	Search table for a target value.
LOOKUP	Look up table and store value in variable.
LOW	Make the output of specified pin low.
NAP	Enter sleep mode for specified period (18 ms to 2.3 sec).
OUTPUT	Make specified pin and output.
PAUSE	Halt program execution for specified time (0 to 65536 ms).
POT	Read potentiometer value (5K to 50K) on specified pin.
PULSIN	Measure an input pulse on specified pin in 10-ms units.
PULSOUT	Generate a pulse by inverting pin for specified time.
PWM	Output pulse-width-modulation on specified pin.
RANDOM	Generate 16-bit random number (0 to 65535).
READ	Read EEPROM location and store value in variable.
RETURN	Return from subroutine.
REVERSE	Reverse I/O of specified pin.
SERIN	Use specified pin to receive serial data.
SEROUT	Use specified pin to transmit serial data.
SLEEP	Enter low-power sleep mode (1 sec to 18 hrs).
SOUND	Output sound on specified pin.
TOGGLE	Toggle output state of specified pin.
WRITE	Write data to EEPROM.

Parts list

- ☐ BASIC Stamps
- ☐ BASIC Stamp editor package
- ☐ H-bridges
- ☐ Servo motors
- ☐ Stepper motors
- ☐ Dc motors
- ☐ Gear boxes

Available from:

Images Company
POB 140742
Staten Island, NY 10314
(718) 698-8305

Speech-controlled mobile robot

7

Speech control facilitates a robot's usefulness. If you think about it, speech is an ideal method for command input. The speech-recognition circuit we will build operates independently from the robot's main intelligence (CPU). This is a good thing because it doesn't require any of the CPU's processing power, except to poll the recognition lines occasionally. Another advantage to this circuit is it's programmable. You program and train the circuit for the words you want it to recognize. This is better than some other chips on the market that use a canned vocabulary. The speech-recognition circuit (SRC) can be easily interfaced to the robot's central processor.

123

Most voice recognition systems on the market today are software programs that require a host computer (usually IBM PC or compatible) and a sound card. The speech recognition system is still software based, even though it requires hardware (sound card). These programs typically run in the background of a DOS or Windows environment, stealing themselves a portion of memory to occupy while allowing another program like Lotus or WordPerfect to run concurrently. The concurrent operation of the speech recognition programs is what slows the operation of any other program that runs using voice recognition.

There are many applications to voice recognition, aside from robotics. Speech recognition will become the method of choice for controlling virtual reality (VR), appliances, toys, tools, and computers. Because of the far-reaching potential of this technology, companies are developing speech recognition. The ability to control and command by speaking directly to a computer (or appliance) will make it easier and increase the efficiency and effectiveness of working with that device.

At its most basic level, a speech-controlled device allows the user to perform parallel tasks, (i.e., hands and eyes are busy elsewhere) while continuing to work with the computer or appliance.

Project

There are two construction projects outlined in this chapter. The first project is a speech-recognition board, and the second project interfaces the speech-recognition board to a robot.

The first project is an easy-to-build programmable speech recognition circuit — programmable in that you program up to 40 words you want the circuit to recognize. This circuit allows you to experiment with many facets of speech recognition technology, and the heart of the circuit is a single IC, the HM2007 speech-recognition chip. It provides the options of recognizing either .96-second or 1.92-second word lengths.

Using the .96 second word length enables the chip to recognize 40 independent words using an $8K \times 8$ static RAM. You have the option to switch to the longer 1.92 second word length. While this reduces the word recognition count to 20 words, the longer length can be used for phrases instead of isolated words. The circuit as described in this article uses the .96 word length, 40-word recognition library.

The chip can operate in a manual mode or in a CPU mode, and the CPU mode is designed to allow the chip to work under a host computer. The circuit we are building operates in the manual mode. The full data sheets on the chip are available for those of you who wish to interface the chip to a computer.

Learning to listen

We take our speech-recognition abilities for granted. Listening to one person speak among several at a party is beyond the capabilities of today's speech-recognition systems, and speech-recognition systems like ours have a hard time separating and filtering out extraneous noise.

Because of this, when using the SRC on a mobile robot, we incorporate two small walkie-talkies. The output of one walkie-talkie is connected to the speech input of the SRC. The other walkie-talkie is used to speak to the robot, and this setup eliminates extraneous noise.

Speech recognition is not speech understanding. Just because a computer can respond to a vocal command, that does not mean the computer understands the command spoken. Future voice-recognition systems will have the ability to distinguish nuances and the meaning of words to "Do what I mean, not what I say!" However, those systems are still years away from being developed.

Speaker dependent/speaker independent

Speech recognition is classified into two categories: speaker dependent and speaker independent. Speaker-dependent systems are trained by the individual who will be using the system. These systems are capable of achieving a high command count and better than 95% accuracy for word recognition. The drawback to this approach is that the system responds accurately only to the individual who trained the system. This is the most common approach employed in software for personal computers.

Speaker-independent systems are trained to respond to a word, regardless of who speaks. Therefore the system must respond to a large variety of speech patterns, inflections, and enunciations of the target word. In speaker-independent systems, the command word count is usually lower than in speaker-dependent systems; however, high accuracy can still be maintained within processing limits. Industrial requirements more often need speaker-independent voice systems.

Our speech recognition circuit is speaker dependent. We can build a little speaker independency by allocating more than one word space to a target word, then programming different word enunciations in the allocated spaces. Each of these word spaces would trigger the same command.

Recognition style

Speech-recognition systems have another constraint concerning the style of speech they can recognize. They assume three styles of speech: isolated, connected, and continuous.

Isolated

Isolated speech-recognition systems can just handle words that are spoken separately, and these are the most common speech-recognition systems available today. The user must pause between each word or command spoken. Our speech-recognition circuit uses isolated words.

Connected

Connected speech recognition is a halfway point between isolated word and continuous speech recognition. Connected speech allows users to speak multiple words. The HM2007 can be set up to identify words or phrases that are 1.92 seconds in length, but this reduces the word-recognition dictionary number to 20.

Continuous

Continuous speech is the natural, conversational speech we are used to hearing in everyday life. It is extremely difficult for a recognizer to shift through the text, as the words tend to merge together. For instance, "Hi, how are you doing?" sounds like "Hi, howyadoin." Continuous speech-recognition systems are on the market and are under continual development.

Speech-recognition circuit

The speech-recognition demonstration circuit operates in the HM2007's manual mode. This mode uses a simple keypad and microphone to program the HM2007 chip.

Keyboard

The keyboard is a telephone keypad made up of 12 normally open switches.

1	2	3
4	5	6
7	8	9
Clear	0	Train

When the circuit is turned on, the HM2007 checks the onboard static RAM. If the RAM checks out okay, the board displays "00" on the segmented displays, lights the LED (READY), and waits for a command.

To train

Press "01" (display will show "01"), then press "T" (Training). Hold the microphone close to your mouth, and say the training word. If the circuit accepted the input, the LED will flash. The word entered is now programmed as the "01" word. If the LED did not flash, start over by pressing "01" then "T." Continue training new words in the circuit. Press "02" then "T" to train the second word. The circuit will accept up to 40 words.

Testing recognition

Repeat a trained word into the microphone. The number of the word should be displayed on the segmented display. For instance, if the word "directory" was trained as word number 25, saying the word "directory" into the microphone will cause the number 25 to be displayed.

Error codes

The chip provides the following error codes.

- ☐ 55 = too long
- ☐ 66 = too short
- ☐ 77 = no match

Clearing memory

To erase RAM memory (Training), press "99" then "CLR."

More about the HM2007 chip

The HM2007 is a single-chip CMOS voice-recognition LSI (large scale integration) circuit. The chip contains an analog front end, voice analysis, recognition, and system control functions. The chip may be used in a stand-alone or connected CPU.

Features

- ☐ Single chip voice recognition CMOS LSI.
- ☐ Speaker-dependent.
- ☐ External RAM support.
- ☐ Maximum 40-word recognition.
- ☐ Maximum word length 1.92 seconds.
- ☐ Microphone support.
- ☐ Manual and CPU modes available.
- ☐ Response time less than 300 milliseconds.
- ☐ 5V power supply.

Circuit construction

The SRC is available in kit form from Images Company (see the parts list at the end of the chapter). The schematic is shown in Fig. 7.1, and the components can be mounted and wired on a standard PCB board.

Solder the keypad to the board according to Fig. 7.2. You will have just seven wires from the keypad to the HM2007 on the PCB. The number next to the wires coming out of the keypad refers to the

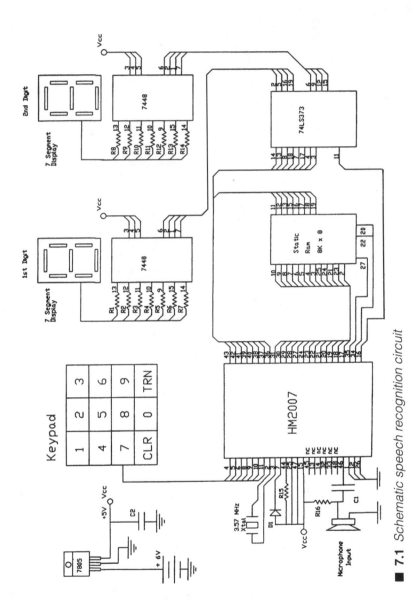

■ 7.1 Schematic speech recognition circuit

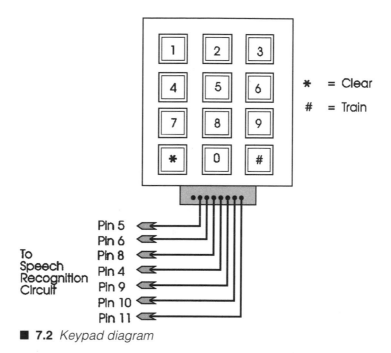

* = Clear

\# = Train

To Speech Recognition Circuit

Pin 5
Pin 6
Pin 8
Pin 4
Pin 9
Pin 10
Pin 11

■ **7.2** *Keypad diagram*

pin number it's connected to on the HM2007 IC. Figure 7.3 shows the top view of the parts placement on the PC board, and Fig. 7.4 is the complete speech-recognition circuit.

Independent recognition system

This demo circuit allows you to experiment with dependent and independent systems. The system is typically trained as speaker dependent, meaning that the voice that trained the circuit also uses it.

We will take the other track and train the system for speaker-independent recognition. To accomplish this, we use four word spaces for each target word.

To simplify the digital logic, the allocation of word spaces is as follows. Our circuit will only look at the first digit space for recognition (least significant digit on the display). This means that the word spaces "01," "11," "21," and "31" will all be recognized as the same word. Since we are only decoding the first digit, they all look like word space "1." Likewise word spaces "04," "14," "24," and "34" all look like word space number 4.

This system works most of the time, but a problem is encountered when an error code pops up:

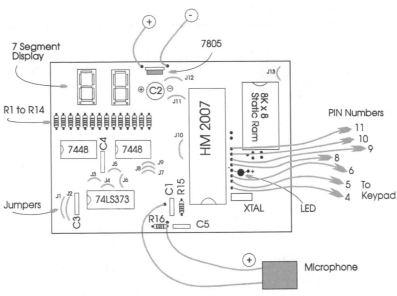

■ **7.3** *Top view of board for parts placement*

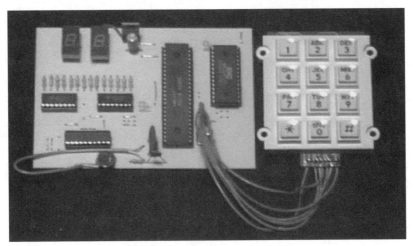

■ **7.4** *Complete speech recognition board*

☐ 55 = word too long

☐ 66 = word too short

☐ 77 = no match

Obviously, the base circuit would identify these error codes as words 5, 6, and 7. There are two work-arounds to this problem. The first work-around is a dedicated circuit (see Fig. 7.5) that brings a line high when the digits 5, 6, or 7 appear in the most significant byte MSB. This line becomes an enable-disable line. The circuit should interpret this line going high as a disable.

The second work-around simply uses the mirocontroller to read the entire 8-bit output; any word number above 40 is ignored. While we are not interfacing this circuit to the BASIC Stamp, it should be evident to anyone who has worked with the Stamp in other applications that this interface would not present a problem.

The 8-bit output is taken from the output of the 74LS373 data octal latch. The output is not a standard 8-bit byte, rather it is broken into two 4-bit binary coded decimal (BCD) nibbles. BCD is related to standard binary numbers as Table 7.1 illustrates.

As you can see, binary and BCD remain the same until it reaches decimal 10. At decimal 10, BCD jumps to the upper nibble, and the lower nibble resets to zero. Binary continues to decimal 15, then jumps to the upper nibble at 16, where the lower nibble resets. If a computer is expecting to read an 8-bit binary number and BCD is provided, this will cause errors.

Interface circuit

The interface circuit revolves around the 4028 BCD integrated circuit. The 4028 takes the BCD output from the 74LS373 on the speech recognition board and outputs a high signal, see truth table on 4028.

The schematic for the interface circuit is shown in Fig. 7.6. The robot car has just four functions: forward, reverse, left, and right. For this, use the Q1–Q4 outputs. Note that the left and right signals must activate both the forward motor and appropriate turn.

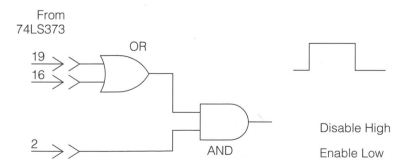

Error Code Detection from MSB of 74LS373

■ **7.5** *High disable from MSB of 74LS373*

■ Table 7.1

Decimal	Binary	BCD
0	0000	0000
1	0001	0001
2	0010	0010
3	0011	0011
4	0100	0100
5	0101	0101
6	0110	0110
7	0111	0111
8	1000	1000
9	1001	1001
10	1010	0001 0000
11	1011	0001 0001
12	1100	0001 0010
13	1101	0001 0011
14	1110	0001 0100
15	1111	0001 0101
16	0001 0000	0001 0110
17	0001 0001	0001 0111
18	0001 0010	0001 1000
19	0001 0011	0001 1001
20	0001 0100	0010 0000

132

Walkie-talkie

The output of the walkie-talkie is connected to pin 46 on the HM2007 through capacitor C1. The capacitor C1 will block any dc component output from the walkie-talkie. As stated previously, the walkie-talkie is used to eliminate noise.

Training and controlling the mobile robot

The speech-recognition board should be trained using the walkie-talkie. Allocate four word spaces for each target word. The digital display on the board is still activated when the interface board is connected, so the display can be used to check recognition accuracy. Find out the range of the walkie-talkie system. Don't let the mobile robot travel outside of its range, or you will end up running after it, yelling stop, stop, stop in the walkie-talkie. Controlling the robot is as simple as talking to it, and it's pretty impressive to boot.

| Input | | | | Output | | | | | | | | | |
D	C	B	A	Q9	Q8	Q7	Q6	Q5	Q4	Q3	Q2	Q1	Q0
0	0	0	0	0	0	0	0	0	0	0	0	0	1
0	0	0	1	0	0	0	0	0	0	0	0	1	0
0	0	1	0	0	0	0	0	0	0	0	1	0	0
0	0	1	1	0	0	0	0	0	0	1	0	0	0
0	1	0	0	0	0	0	0	0	1	0	0	0	0
0	1	0	1	0	0	0	0	1	0	0	0	0	0
0	1	1	0	0	0	0	1	0	0	0	0	0	0
0	1	1	1	0	0	1	0	0	0	0	0	0	0
1	0	0	0	0	1	0	0	0	0	0	0	0	0
1	0	0	1	1	0	0	0	0	0	0	0	0	0

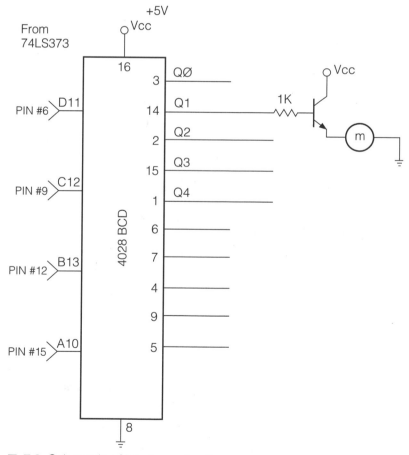

■ **7.6** *Schematic of interface circuit*

133

■ **7.7** *Speech-controlled mobile robot*

New board features

The voice-controlled robot is shown in Fig. 7.7. The circuit board on the robot looks a little different than Fig. 7.4 because I happened to get a prototype of the latest revision of the speech kit.

This revision makes interfacing to the board much easier. There's a 9-pin header (two 4-bit nibbles plus ground) that connects to the 74LS373. In addition, the board has a 3-volt input for memory backup. This makes the static ram on the speechboard nonvolatile. So you can turn the board on and off without losing the words programmed in the static ram. In the original version when you turned off the power you lost the words programmed in the ram. By the time this book goes to press, the newly revised speech recognition board will be available.

Other applications

The interface to the speech board can also be used for controlling ac and dc power (see Fig. 7.8). The speech-recognition board can be interfaced to an electric wheelchair for the disabled. Another idea is to create an IBM-compatible speech-controlled computer keyboard.

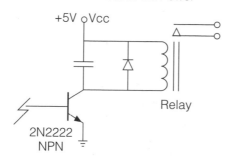

Ac or Dc Power

+5V Vcc

Relay

2N2222
NPN

Ac Power Controller

330Ω   180Ω   Socket

MOC
3010

Triac

■ **7.8** *Controlling ac and dc power using speech circuit*

135

Parts list

QUAN	ITEM	COST
1	IC1 HM2007 IC	$25.00
1	IC2 SRAM 8K X 8	$8.00
1	IC3 74LS373	$2.00
2	IC4 & IC5 7448	$2.00 ea
1	XTAL 3.57 MHz	$3.00
1	PCB	$20.00
1	Keypad	$12.00 ea
2	7 Segment Displays	$3.00 ea
14	R1–R14 220-ohm 1/4-watt resistor	
1	R15 22K 1/4-watt resistor	
1	R16 5.6K 1/4-watt resistor	
3	C1,C3 C4 .047-uf cap	
1	C2 100-uf 16V cap	
1	C5 .1-uf cap	
	Microphone	
	9V Battery Clip	

■ Table 7.3 HM2007 Pin Functions

No.	Name	I/O	Description
1	GND-		Ground
2,3	X1,X2	I	Crystal Pins, A 3.57 crystal is connected to these pins.
4,5,6	S1,S2,S3	I/O	Keypad scanning pins for manual mode.
7	RDY	O	Voice Input Indicator. Active Low. If HM2007 is ready for voice input a low signal is output. If chip is busy, signal is high.
8-11	K1-K4	I/O	Keypad scanning pins for manual mode.
12	Test	I	"H" = Test Mode. "L" = normal mode
13	WLEN	I	Word length select pin. Selects the word length to be recognized. When pulled high (+5V), word length= 1.92 second. Pulled low, word length= .96
14	CPUM	I	Mode select High = CPU mode Low = Manual
15	Wait	I	Wait Low = Wait High = Active
16	DEN	O	Data Enable
17–24	SA0–SA7	O	External Memory Address Bus
25,47	Vdd		+5V
26,48	GND		Ground
2–7-31	SA9-SA12		External Memory Address Bus
32,33	NC		Not Connected
34	ME	O	Memory Enable
35	MR/MW	O	Memory read/Memory write. Pin connects to the R/W pin of the SRAM.
36–43	D0–D7	I/O	External Memory Data Bus
44	Vref	I	Voltage Ref input for internal ADC.
45	Line	O	For testing only.
46	MicIn	I	Microphone connect pin.

Parts available from :
Images Company
POB 140742
Staten Island, NY 10314
(718) 698-8305

For shipping, add $7.50. New York state residents add 8.25% in sales tax. HM2007 data sheets are $5.00. A complete speech recognition kit is available for $100.00 plus $7.50 postage and handling from Images Company.

Neural nets, nervous nets, and subsumption architecture

8

The class of robotics discussed in this chapter does not have a central processing unit (CPU). Rather, it functions on a stimulus-response mechanism, also called subsumption architecture, pioneered by Rodney Brooks at MIT. Different stimulus responses can be layered on top of one another, and a multilayer stimulus-response mechanism can exhibit what appears as intelligent behavior.

Intelligent photovore robot

To illustrate this point, let's build a few layers of stimulus response and see if we can create what may appear as intelligent behavior. In Chapter 6 we programmed a photoresistive light-tracking system, which locks onto a light source and tracks it. If placed on a mobile robot and allowed to steer, the tracking system directs the robot toward a light source. This is our first stimulus-response layer. This type of robot is called a *photovore* or "light eater."

The program illustrates how the rule-based BASIC Stamp can perform neural functions. For example, let's now design a circuit that performs the same function without any "rule-based" intelligence.

Figure 8.1 uses an 8-pin-dip single-voltage supply dual op-amp. The op-amps are configured as comparators (which were covered more fully in Chapter 5). If you have any questions about Fig. 8.1, review Chapter 5. Two CdS photoresistors are wired in series, forming a voltage divider. The output of the photoresistor voltage divider is fed into the inverting input of one op-amp and the non-inverting input of the other op-amp.

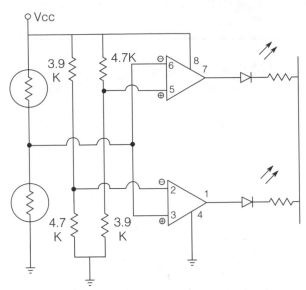

■ **8.1** *Dual op-amp neural comparator circuit*

Two other voltage dividers are needed. They are symmetric but opposite. One voltage divider has a 3.9K-ohm resistor connected to Vcc and a 4.7K-ohm resistor connected to ground. The second voltage divider uses the same value resistors, in opposite positions.

When both photoresistors are evenly illuminated, neither LED is lit. Cover one or the other photoresisitor, and the corresponding LED will light.

Each op-amp is acting like a simple electronic neuron. When the electrical stimulus falls above or below (depending on which op-amp we're talking about) a threshold (determined by the 3.9K and 4.7K voltage divider), the neuron fires.

The firing of the neuron (or outputs of the op-amp) can be used to turn on a dc motor using an NPN transistor (see Fig. 8.2). The dc motors in turn provide movement and direction for the photovore robot.

To create a simple photovore robot, a chassis is designed that has two gear-box dc motors (see Fig. 8.3). When both motors are powered, the robot moves forward in a straight line. To turn left or right, one or the other motor is turned off. The motor that still receives power turns the robot.

For our photovore robot, we need both motors to be powered when the two photoresistors are evenly illuminated. Running the outputs of each op-amp into an inverting buffer before the NPN transistor accomplishes this task (see Fig. 8.4).

Behavior

When one photoresistor receives less light than the other, the corresponding motor will turn off, allowing the motor that's still powered to turn the robot toward the light source. When the robot turns so that both photoresistors are again evenly illuminated,

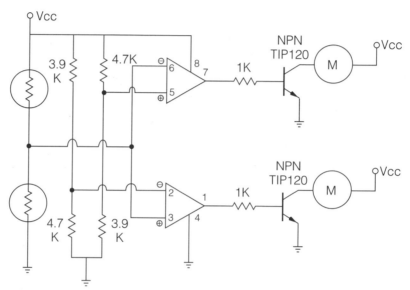

■ **8.2** *Neural comparator dc motor control circuit*

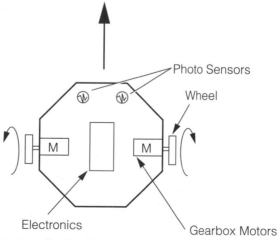

■ **8.3** *Outline diagram of photovore robot*

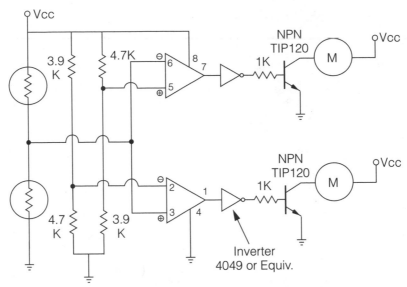

■ **8.4** *Neural comparator dc motor control circuit with inverters*

both motors turn on, allowing the robot to travel toward the light source.

Light avoidance

If we crisscross the outputs of the op-amps to the motors, the behavior reverses. Instead of moving toward a light source, the robot now avoids light and seeks shelter.

Adding behavior (feeding)

We can add behavior to the photovore by adding another stimulus-response layer (see Fig. 8.5). This will be another light-activated comparator circuit that facilitates feeding. (Comparators were covered more fully in Chapter 5. If you have any questions about Fig. 8.5, go back to Chapter 5.) The second layer is placed on top of the first layer. When the light intensity is great enough, this threshold detector cuts power to the first layer and the motor drive system. If we place a number of photovoltaic cells and a diode, the electric power generated from the photovoltaics can trickle charge a nicad power pack (Vcc). Let's call this function feeding.

Still more behavior (resting)

We don't want our photovore traveling around in the dark wasting precious energy. So let's add another layer. The third layer is another light threshold detector, see Fig. 8.6. This detector cuts

2nd Layer

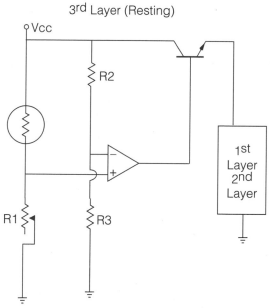

Vcc

R2

R1

R3

First
Layer

To Vcc

Photovoltaics
Trickle
Charger

■ **8.5** *Feeding behavior comparator circuit*

3rd Layer (Resting)

Vcc

R2

R1

R3

1st
Layer
2nd
Layer

■ **8.6** *Resting behavior comparator circuit*

power to the first layer, motor drive system and second layer in darkness or near darkness. When a sufficient amount of ambient light is reintroduced power to first layer, the drive system and second layer are restored.

Emergent behavior

Okay, let's look at the behavior of our three-layer stimulus-response photovore robot and see if we can classify its behavior as intelligent. In complete darkness the robot remains still, conserving all its power via layer 3. As ambient light is introduced and increased, layer 3 restores power to the drive system and the first two layers. At this point, layer 1 takes over and controls the direction of the robot. The robot searches and moves toward the source of light. As the robot moves toward the light source, the light intensity increases. When the light reaches a sufficient intensity, layer 2 cuts power to the drive system, allowing the robot to feed (charge its batteries) through the photovoltaics.

Whether you decide to classify this robotic behavior as intelligent or not is an individual preference and one that can clearly be debated on both sides of the fence. At least it illustrates how complex behavior patterns can be generated using a layered stimulus response.

142

BEAM robotics

BEAM is an multidimensional acronym that loosely stands for biology, electronics, aesthetics, and mechanics. I say loosely because there are numerous groups of words that can be and are at times substituted in the acronym (for instance, biotechnology, evolution, analog, and modularity).

BEAM robotics has an annual olympic competition with 14 events. Later we will build a robotic device that may be entered in the solar roller competition. But first let's learn a little more about BEAM robotics.

BEAM robotics was founded by Mark Tilden while at the University of Waterloo in Canada. The inspiration for BEAM-style robots came from a talk given by Rodney Brooks of MIT that Mark had attended in 1989. Dr. Brook's approach to robotics removes the standard onboard intelligence, sometimes refered to as "CPU," "rule base," "expert system," or "program" from the robot. Instead, the robot performs useful functions by hard-wiring particular stimulus responses.

BEAM competition

The first BEAM competition was held in 1991. The inspiration for the BEAM games came from the first international Robot Olympics held in Glasgow, Scotland, in 1990. A central idea to the BEAM philosophy is robotic evolution. Start simple and evolve toward complex systems. As illustrated, the idea is to break away from standard robotic design, using top-heavy CPUs for control and embracing a bottom-up approach using a layered stimulus response (neural network, nervous network system). Mark Tilden calls his stimulus-response mechanisms nervous nets.

Mark has designed a number of interesting robots (see Fig. 8.7). They employ what he calls a nervous net system that is made using transistors. Since the nervous net system is patented (by Mark Tilden), unpublished schematics for his nervous net system are not readily available. I do not have anything available to look at this nervous net system. However, Mark has a book in the works titled *Living Machines*, which is planned for publication sometime in 1997. In this book, schematics for the nervous net on Mark's first robots will be published. In addition to *Living Machines*, Mark

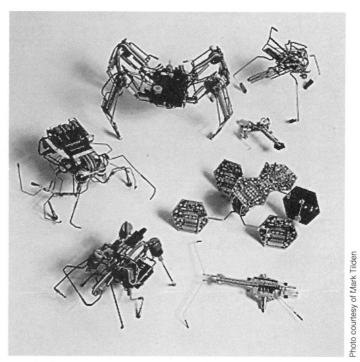

Photo courtesy of Mark Tilden

■ **8.7** *BEAM robots*

Tilden is writing a second book titled *Robotica*, where the nervous net technology will be fully explored.

To get an idea of the type of BEAM robots Mark builds, let's look at a few designs. Figure 8.8 is titled Newspotter. It is an eight-transistor solar-powered device designed to study the nature of

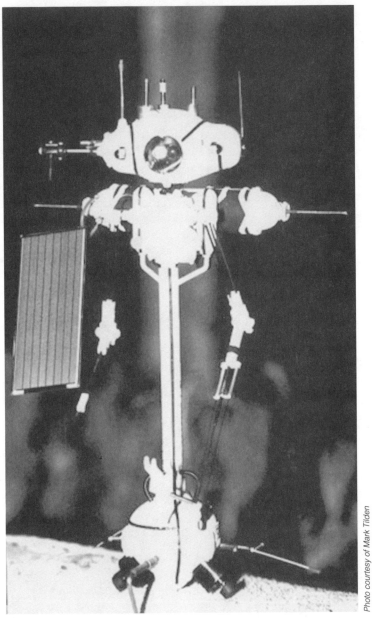

Photo courtesy of Mark Tilden

■ **8.8** *Newspotter robot.*

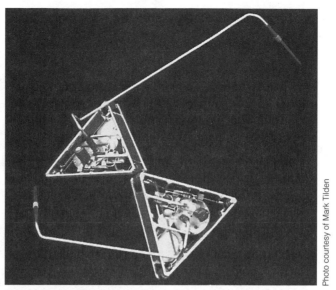

Photo courtesy of Mark Tilden

■ **8.9** *Tilebots*

head motion with respect to wheeled action. The device is 8 inches high and was designed as a prospective NASA prototype.

Figure 8.9 is titled Tilebots. Mark Tilden asks, "What would a triangle do if it could?" The tilebots are single-neuron prototypes showing that devices of like structure can self-assemble themselves into larger living machines. Each device is solar powered and 3 inches across.

Figure 8.10 is titled Gumby Trks. This is a type of biomechanical walker that is being designed for a variety of terrains. Here Gumby 1.0, an 8-transistor imbedded-bicore walker about a foot long makes tracks across a sand desert.

Figure 8.11 is titled Walkman 1.0. The first of the 12-transistor "microcore" walkers, this device was put together from the remains of five similar Walkman cassette players. It has seven sensors, including two eyes, and can handle very complex terrains with its five-motor design.

Electronic flotsam

BEAM robotists pride themselves on using discarded electronics in the construction of their robots—for instance, solar cells from calculators, high-efficiency electric motors from Walkmans and cassette players, along with pulleys, switches, capacitors, components, gears, and solenoids. Gathering this electronic flotsam and converting it into useful robots is a project in recycling engineering.

Neural nets, nervous nets, and subsumption architecture

■ **8.10** *Gumby Trks.*

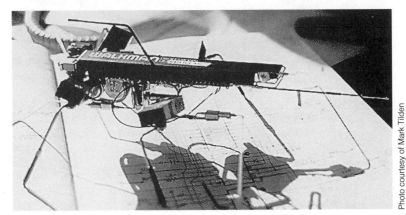

■ **8.11** *Walkman 1.0*

The BEAM robotic competition is open to everyone, and all competitors start on equal footing. A seven-year-old robotist has as much chance of winning as a professor from a prestigious college. In some cases, a seven-year-old won!

Competitions

The following is a brief synopsis of the competitions held at BEAM Games. Complete descriptions of the events and rules can be found in the BEAM guide available from the University of California. The address is listed at the end of this chapter.

Solar roller

Create a solar-powered robot racer. Maximum robot size—must fit into a 6-inch cube. Maximum solar cell size is $\frac{1}{2}" \times 2\frac{1}{2}"$ (1.25 square inches). Track length is 1 meter, width 6 inches. Competitors race in full sunlight (or 500-watt halogen lamp equivalent).

☐ Class A: Race on level sheet of glass.

☐ Class B: Rough terrain.

Photovore

Create a solar-powered goal-seeking robot whose overall size can fit within a 7-inch cube. The robot will be placed with other competitors in a closed "Jurassic Park" for 30 hours. Those robots that show the best survival, exploration, confrontation, speed, and power efficiency (determined by review of photos and video) will be the winners.

Aquavore

Create a solar-powered robot whose overall size can fit inside a 7-inch cube that will swim the length of a 55-gallon fish tank (distance approximately one meter). A 6-inch-high wall will be placed halfway in the tank, and the competitor must pass the wall to reach the finish line.

Robot limbo

Create a robot whose overall size can fit inside a 7-inch cube that will run through a simple maze. Solar power is not required for this competition, but is recommended.

Robot rope climbing

Create a robot that climbs up a meter of rope and then back down. The fastest wins. The rope is 40-lb test nylon fishing line. Overall size of the robot must fit within a 20-inch cube.

Robot high/long jump

A class Create a robot that can jump with its entire mass into the air three times using the power from one optional battery. Robot must fit within a 1-square-foot space.

B class Create a robot that can jump with its entire mass forward three times using the power from one optional battery. Robot must fit within a 1-square-foot space.

Legged robots

Legged robots compete with each other. They are given points based on their ability to walk over various terrains and negotiate obstacles. There is no size restriction.

Innovation machines

Create a new device, the purpose of which need not be obvious. Competitors are judged on the quality of workmanship, broadness of scope, and weirdness of application.

Robot art: best-modified appliance competition

Create a robot that can draw or generate art. The generation of art may be the movement of the robot itself. An example given is a solar flower that opens slowly and snaps closed when light shines upon it.

☐ Class A: Robots built completely from scratch.

☐ Class B: Modified devices, toys, appliances, etc.

Robot sumo wrestling

Class A Robots are paired together in competition. Each robot attempts to push the other off the edge of a 5-foot-round platform. Robots can be self-contained, tethered, or RC.

Class B Robots try to push each other off a 6-foot-round platform.

Nanomouse competition

Create a self-contained robotic mouse that can run through a maze. The robot's footprint must be no larger the 10 centimeters × 10 centimeters. There is no restriction on height.

Micromouse competition

Create a self-contained robotic mouse that can run through a maze. The robot's footprint must be no larger than 25 cm × 25 cm. There is no restriction on height.

Aerobot competition

Create a flying robot that will launch itself, fly into a 25-foot-by-25-foot drop zone, find a randomly placed target in the drop zone, drop a marker on it, and then return to its launch pad.

Miscellaneous competitions

If you have built a robot that doesn't fit in the outlined categories, it may be entered in the miscellaneous category.

Getting the BEAM guide

Complete 120-page BEAM guides may be purchased for $20.00. Make checks payable to the University of California: BEAM Games. For current information, call or write:

BEAM Robot Olympics c/o Mark W. Tilden
Mail Stop D449
Los Alamos National Labs
Los Alamos, NM 87545
(505) 667-2902

The Internet address for BEAM Games is:

http://sst.lanl.gov/robot/

Join in

The BEAM competitions are open to all robotists. You can enter a robot in the competition or just attend the event for fun. Contact the BEAM Robotic Olympics (address given above) to get up-to-date information. The following are plans to build a simple solar-roller robot.

Solar engine

The solar engine is commonly used as an onboard power plant for BEAM-type robots, sometimes called living robots. We will use the solar engine circuit described in Chapter 3.

Roller

Those of you who are mechanically inclined may want to build your own lightweight chassis. If you're like me, the best thing to do is to find a toy car chassis that will lend itself to a simple modification.

I found a small plastic toy car, cut the top of the plastic car off using small cutting pliers, and fit the solar-engine PC board inside the car chassis. The solar cell is positioned on top of the PC board (see Fig. 8.12).

The most difficult part—creating a drive system—turned out to have a simple solution. The metal shaft of the HE motor doesn't provide a good traction or friction surface, so the first thing to do is to find some small-diameter rubber tubing and secure it onto the metal shaft. If you have large-gauge wire around, you may be able to strip an inch of insulation off and use that on the motor shaft. On the prototype I built, I had some leftover rubber covers for microtoggle switches. These fit on the shaft perfectly.

The motor is mounted in the chassis so that the motor's covered shaft leans on the back wheel. When the circuit discharges, the shaft spins the wheel, moving the car forward. Granted, this simple friction drive is not very efficient, but it is simple and it does work. Feel free to create your own (more efficient) drive mechanism.

■ **8.12** *Solar Roller. Toy car using solar engine*

Expanding games

The success of the BEAM competitions are spawning new games like the West Canadian Games. This and other useful links can be found on the BEAM home page listed. For the solar engine parts list, see Chapter 3.

Audio graffiti

Let's build an autonomous robotic message system. Imagine solar-powered robots littering the landscape. Pick one up and listen to a message someone left. You can either leave the existing message or record a new one as you see fit. What would you call this type of autonomous message system? The best description I came up with is audio graffiti.

Now, for this scheme to work, the solar-powered robots must be cheap, reliable, and weatherproof. While the robot outlined here does not meet all the criteria, it is a step, an evolutionary step you could say, in the right direction.

Message system

There are a number of digital message ICs around. The support circuitry required is neither complex nor expensive. Even so, there are stand-alone boxes one can purchase that are cheaper still and don't have to be assembled. You've seen one on TV. It's called the Yak-Bak (TM) (see Fig. 8.13). Now you don't need the last iteration of the Yak-Bak with its sound-morphing features. The original model that records and plays back straight sound works well. However, if you want the sound-morphing feature, and there's no reason not to just get a later edition, the conversion should work just as well.

The Yak-Bak has all the electronic message circuitry we need. Begin by opening up the case (see Fig. 8.14) and removing the printed circuit board. The microphone and speaker are attached to the board and should remain so. Disconnect the button cell batteries.

Solar powered

To make the Yak-Bak solar powered requires two solar cells and three high-capacity capacitors (see Fig. 8.15). The capacitors used in the prototype are .047 farads. The solar cells used are rated at 1.6 volts each. However, each solar cell will provide 2.5V or better under no-load or low-load conditions. Placing two solar cells in series can provide +5 volts to the capacitor bank.

■ **8.13** *Yak-Bak*

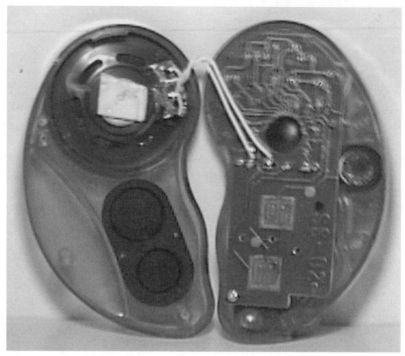

■ **8.14** *Opening Yak base case to reveal electronics*

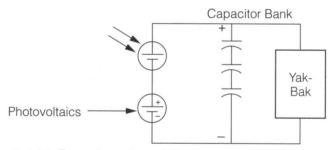

■ **8.15** *Two solar cells and three capacitors make up the power supply*

The capacitors are strung in parallel. This provides the greatest electrical storage capacity ($3 \times .047$ farad = $.141$ farad) for the electrical circuit. It takes a while for the solar cells to charge the capacitor bank, but once charged, the Yak-Bak will function for quite a while on one charge.

Maintaining the message through the night

The Yak-Bak requires a nominal voltage to keep a message stored in memory. I've tested the unit down to 2.5 volts, and the message is still retained. During the night, no sunlight falls on the cells, and the capacitor bank begins to deplete during darkness. After 12 straight hours of darkness, the unit maintains its memory and can still function. The capacitor bank charge dropped about 1.5V (to 3.5V) during the dark period.

Next step

I placed all the loose components (PC board, capacitors, solar cells, etc.) on a sheet of transparent acrylic plastic (see Fig. 8.16). I was not happy with the overall appearance, and I believe that the next evolutionary step would be to keep the original Yak-Bak case.

When building the prototype, the reason I didn't put the Yak-Bak circuitry back into its original housing was that the capacitors would not fit in. I think the next step for someone wanting to improve the design would be to find high-capacity capacitors that will fit inside the housing, taking the space originally occupied by the button batteries. This will make for a much cleaner-looking unit. The solar cells will still need to be on the outside, but they can be arranged in a decorative, functional fashion (see Fig. 8.17).

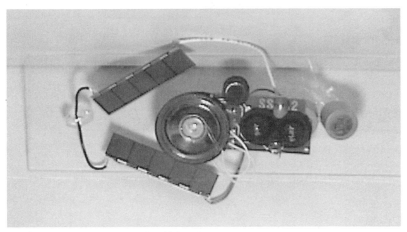

■ **8.16** *Finished Audio Graffiti unit*

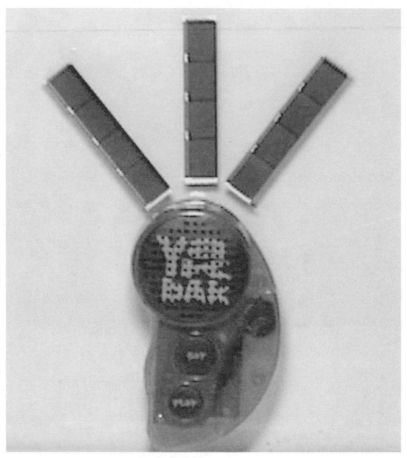

■ **8.17** *Mock-up of a more advanced Audio Graffiti unit*

Parts list

Quan	Item	
1	Yak-Bak call	
2	Solar Cells SC-01	$3.00 each
3	.047-farad capacitors	$2.50 each

Images Company
POB 140742
Staten Island, NY 10314
(718) 698-8305
(718) 982-6145 FAX

The Internet site is:

http://www.imagesco.com

Telepresence robots

9

In this chapter we will build a telepresence robot (T-bot). Telepresence robots are planned for use in a variety of science, entertainment, business, military, exploratory, and industrial applications as illustrated in Chapter 2.

What's in a name?

The late science fiction (SF) writer Robert Heinlein is credited as the first person to predict the use of telepresence robots in his 1940 SF novel titled *Waldo*. In the story, a human operates mechanical puppets, called "waldos," to do his bidding from a remote location.

Rather than use the term "waldo" or "avartar," I found the word "golem" from Yiddish mythology more suitable to define a physical telepresence system. (More recently, the word "avartar" is used to define an online presence, or presence in cyberspace.) The story of the golem describes the intentional placing of a human spirit in a clay figure. The spirit controls the clay figure to do its bidding, which it would not or could not do in its human form. Once the golem's work is finished, the spirit returns to its human form. This definition adequately describes the new science of telepresence. I therefore have named my first telepresence robot Golem I.

What is telepresence?

Telepresence is a high fidelity form of remote control that projects the senses of the human operator into a robot at a distant site. The interfaces used to create a telepresence system are the same as used in VR. Figure 9.1 illustrates a basic telepresence system.

In virtual reality we achieve immersion into a synthetic computer generated environment by fooling our senses, as best we can, to believe in and interact with the computer synthesized environment. In telepresence the environment is real but remote, so instead of a

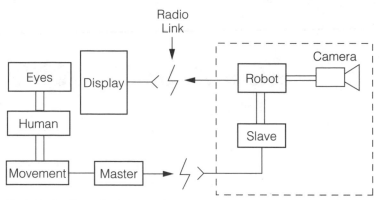

■ **9.1** *Outline of a basic telepresence system*

computer generating a synthetic environment, the sensors placed on the remote robot feed all spatial and environmental information.

On the human side, as stated before, the same VR equipment is used to fool our senses into believing that the environment is real. In this case, however, our immersion (presence) is at the remote location of the robot.

Different levels of presence are achieved depending upon the fidelity of the interfacing devices. A humanoid robot that could accurately follow human movement, gestures, locomotion, and balance while providing visual, thermal, tactile, and force reflection over its entire exo-skeleton to the human operator would be a perfect golem. The illusion created is that the operator has merged or is contained within the robot structure.

Current telepresence systems fall quite short of this goal. In most cases, however, the remote robot is a vehicle, like the one we shall build. The best telepresence existence available with these rudimentary T-bots is to believe one is actually driving the vehicle from inside. This is the telepresence we are striving for.

T-bots can be built to explore and operate in harsh or hazardous environments. A partial list of remote environments includes arctic underwater diving, ocean-floor rover, forest fire, active volcanoes, nuclear reactor, the Moon, Mars, or anything in between.

System substructure

The framework upon which we will build our T-bot is a radio-controlled (RC) electric car. Ideally, the model car should have proportional steering and speed control. This is the type of model used in building the prototype. A less expensive RC model may be used, but you won't have as much control when driving.

Figure 9.2 is a photograph of the RC model car. It has a spring suspension system that can be incorporated with a sensor system to provide a feel of the terrain. But we are getting ahead of ourselves.

Purchase an RC car that is bundled with a battery charger and rechargeable batteries. With some RC models, these items must be purchased separately.

A little on RC models

Radio-controlled models have evolved into a popular hobby. There are radio-controlled (RC) airplanes, helicopters, gliders, power boats, submarines, cars, motorcycles, etc. Most models are suitable shells and springboards for golem-type robots.

Not long ago, RC models were exclusively gas powered. In the late 1970s, improvements in battery technology and electric motors made electric-powered vehicles a viable option.

Model cars are typically controlled from a two-channel radio. One channel controls steering and the other throttle. Each channel is controlled by a potentiometer inside the transmitter. The steering potentiometer is often connected to a small steering wheel on the transmitter control. The throttle is usually connected to a trigger or stick.

■ **9.2** *RC model car used in telepresence system*

An encoder chip in the transmitter modulates a pulse width on the transmitter's carrier signal. The pulse width is based on the position (resistance) of the potentiometer's shaft. The pulse widths are varied between 1 and 2 milliseconds (ms) (see Fig. 9.3). When the potentiometer is in its center position, the pulse width corresponding to that channel is 1.5 milliseconds. When the control is pushed to one extreme, the pulse width increases to 2 ms. When pushed to the opposite extreme, the pulse-width shrinks to 1 ms.

The receiver decodes the pulses on the carrier signal and sends them to their respective servomotors. The servomotor is an integral unit, containing a motor, gearbox, output shaft, and a PC board. The PC board on the inside the servomotor generates a reference pulse that is based on the position of an internal potentiometer connected to the output shaft. A decoder chip on the internal PC board compares the incoming pulses from the receiver to the reference pulses. The servomotor attempts to match the pulse-widths of the two signals by adjusting the position of the servomotor's output shaft. This is how the servomotor tracks and holds its position based on the signal from the transmitter.

Eyes

The eye(s) for our T-bot is a microminiature video camera, see Fig. 9.4. The B/W camera cost is approximately $85.00. Color versions are available for $100 more and are well worth the additional money if your pocket-book can afford it.

The overall size of the camera is quite small, so small, in fact, that two video cameras may be mounted side by side and have the approximate IPD of 63 mm between lenses. Mounting a pair of cameras like this will enable the T-bot to transmit realistic stereo pictures to the operator. For the prototype we will use just one camera; later we will discuss improvements to the system that will add stereo-vision to provide depth perception to improve telepresence and operation.

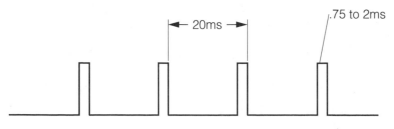

Pulse Signal to Servo Motor

■ **9.3** *Pulse width used to control the servo motor*

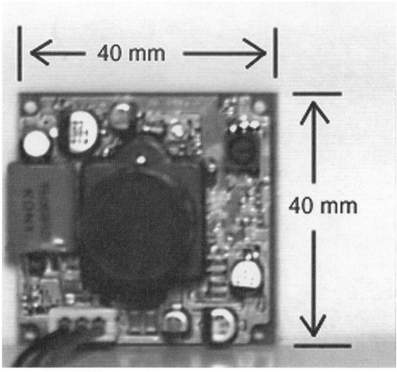

■ **9.4** *Miniature CCD video camera*

The prototype T-bot contains a single microminiature video camera. As with the HMD, the T-bot subsystems are built in modular form. Therefore, if the reader wishes to create a stereo system in the future for the T-bot, all video components are reusable.

Construction

Construction of the T-bot begins by analyzing the chassis of the model car. Most RC model cars have an outer cosmetic shell that makes the model look like a standard vehicle: car, truck, land-rover, etc. Remove the outer cosmetic shell of the RC model and secure the equipment directly onto the chassis.

The Golem I needs three separate power supplies: one each for the video camera, video/sound transmitter, and servo motor. Both camera and transmitter each use a 12-volt battery power supply made from eight AA batteries (see Fig. 9.5). The servomotor with receiver module uses a 6-volt power supply. Although it may be possible to supply both the camera and receiver from a single 12V

■ **9.5** *12-volt battery power supply*

supply, this will deplete the batteries too quickly, making the T-bot run time short.

When purchasing the AA battery compartments, make sure they are the type that have the terminals for a standard 9V snap-on battery clip. This will reduce the amount of soldered wires and simplify construction.

To keep the component mounting simple and modular we will make liberal use of Velcro material. Velcro material is typically sold in strips, by foot increments. The velcro strip is made of two mating strips of material that adhere to one another. Each strip has self-adhesive on the back. The velcro material is mounted to the chassis with the mating strip placed on the component to be mounted.

The battery supplies are mounted to the car using Velcro. This allows for easy removal of the battery packs from the car to replace worn out batteries. Velcro is used on top of the battery compartments for mounting the video transmitter (see Fig. 9.5).

Video transmitter

The video transmitter is a kit available from Images Co. (see Fig. 9.6). Put the kit together according to the instructions supplied. Before tuning the transmitter to channel 3 or 4, we need to make simple modifications to improve its performance.

The first modification on the video transmitter is the power supply. The transmitter is supposed to run using a single 9-volt battery, but a single 9-volt battery power supply is too weak to operate the transmitter to specification. Instead, use a 12-volt battery supply made from 8 AA batteries.

The second modification calls for placing a 75-ohm $1/4$ watt resistor on the transmitter output. The 75-ohm resistor is placed on the solder side of the PC board. Solder one end of the resistor to the antenna pad and the other end to a circuit ground pad. Having made these two modifications outlined above, tune the transmitter according to the directions supplied with the kit.

The transmitter will transmit video and sound to a standard television set. Keep in mind that, for it to work, the TV cannot be connected to a cable box. The TV must use its antenna and tuner.

Mounting the video transmitter

In the prototype, the video transmitter is secured on top of the batteries using Velcro. The 9-volt battery holder on the transmitter PC board is not needed to hold a battery.

■ **9.6** *Video transmitter kit*

Miniature video camera

The miniature video camera does not require construction. It is a finished product and may be used as it is received. The camera has three wires coming out of it to connect power and output the video signal (see Fig. 9.7). The experimenter needs to supply power to the camera and make a suitable mounting bracket to hold the camera onto the Golem robot. The video-out from the camera goes directly to the video-in of the video transmitter.

Mounting the video camera

There are two options regarding the mounting of the video camera. You can start out with a fixed camera position, or it can be mounted on top of a servomotor for head tracking. The servomotor is naturally more complex and requires the building of a head-tracking unit. You may want to try the fixed position first, to keep the construction simple.

Fixed mounting

You need a small bracket to secure the camera to the chassis. The corners of the camera's PC board have four small holes suitable for mounting the camera see Fig. 9.8. The finished Golem robot is illustrated in Fig. 9.9.

Driving via telepresence

The TV transmitter has been tuned to transmit on either channel 3 or channel 4, depending on which TV channel is unused in your area. With the TV set on and receiving on the channel the transmitter is transmitting on, you should be receiving the video from the camera mounted on the RC car. You can drive the car remotely using the RC while looking at the TV monitor.

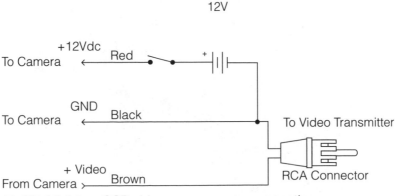

■ **9.7** *Miniature CCD video camera power connection*

■ **9.8** *Video camera mounting*

■ **9.9** *The finished Golem robot*

Adding realistic car controls

Golem I uses the standard RC control that came with the car. The realism of the telepresence system can be greatly improved by adding realistic car controls. This isn't difficult. What is involved is removing the circuit and potentiometers from the RC transmitter control, making a mock-up steering wheel for one potentiometer, and building a foot-pedal throttle control for the other potentiometer.

Adding sound

Golem I has the facilities for sound. All you need do is add a microphone and preamp to the car. The signal output from the preamp is fed in the sound-in RCA plug on the video transmitter.

The sound will be transmitted to the TV along with the video. If you wish, you can plug headphones into the earphone jack on the LCD TV for private listening.

Talk

A child's walkie-talkie can be placed on the T-bot allowing the operator to speak from the T-bot.

Improving the telepresence System

Golem I is a basic telepresence system that can be improved tremendously with a little thought. All improvements will add cost to the system. However, these subsystems may be added over a period of time.

Stereo vision

Implementing stereo-vision on the Golem I is a worthwhile endeavor. There are a great many benefits to be derived, most importantly depth perception. Before undertaking this project it is important to realize that a head mounted display (HMD) will be needed to view these images.

The small size of the miniature video camera is perfect for stereographic imaging. Two cameras can be positioned side by side at the same interocular distance (IOD) of the eyes. To be more specific, the average interocular distance for adult humans is 63 mm. The cameras lens can be positioned, from center to center, at 63 mm apart. This is the same space between the lenses that the eyes use for depth perception.

Seeing in stereo from the Golem provides depth perception to the operators when they are driving. Stereo-vision becomes increasingly more important when robotic arms are in use. Being able to see the manipulator (robot arm) move along the Z axis allows for efficient operation.

The Z dimension is lost in a monocular view. This forces operators to gently bump into objects to approximate the robotic manipulator's location along the Z axis.

In addition to the video cameras, two video transmitters, one for each camera are required. The transmitters should be tuned to transmit on different TV channels.

When the stereo system is set up, the operator of the vehicle will see the T-bot's environment as a three-dimensional picture. However, the stereo images transmitted will not contain convergence clues. The stereo video cameras are in a fixed position looking straight ahead. To include convergence clues would require extensive engineering. A feed forward HMD would need to incorporate convergent eye tracking. The eye tracking information would then be transmitted to servomotors that would rotate the video cameras in direct proportion to the eye convergence. A master-slave system such as this is currently beyond the hobbyist market.

Digital compass

Chapter 5 includes plans for a digital compass that is suitable for use of the Golem. The compass can be set up in two different ways. The first method keeps the LEDs of the compass in the visual field of the video camera. A quick look informs the operator in which direction the Golem is traveling. The second option uses a radio link between the digital output of the compass on the T-bot and the remote location of the operator.

The digital compass is needed more on boat, submarine, or flying golems discussed in the next chapter. Golem I can be used to for systems design and checkout.

Rumble interface

When driving the model car via telepresence, you cannot feel roughness of the road as you drive. To incorporate a rumble feature into the system you could use the spring suspension of the model car. Any number of sensors (such as piezoelectric transducers, hall devices, and strain gauges) can be used for this purpose.

The challenge to the experimenter is not in detecting the rumble but in providing that information to the seat of the operator. Most motion platforms use expensive pneumatic and hydraulic systems. If cost is an issue, this isn't an option.

A cheaper solution can be found in the ThunderSeat by Thunder-Seat Technologies. The ThunderSeat uses any sound source to generate vibratory sensations. It contains a subwoofer speaker coupled to an acoustical wave chamber inside the seat. The wave chamber vibrates the entire seat. The low frequency (woofer) speaker can handle up to 100 watts of power. Frequency response of the system is 50 Hz–3.7 kHz. Originally designed to work with flight-simulator programs running with a sound card on a PC, the output from the sound card is fed into an amplifier then to the ThunderSeat.

Tilt interface

As with the rumble interface there are several transducers one can use to determine tilt (see Chapter 5). One tilt sensor uses a steel ball in a plastic enclosure. When tilted, the steel ball makes contact with electrodes placed in the enclosure. Mercury switches may also be used.

Electrolytic tilt sensors are expensive but are excellent sensors. A single electrolytic sensor can provide tilt information from two axes. The hermetically sealed sensor has one center electrode surrounded by four equidistance electrodes. As the electrolytic fluid makes contact with the internal electrodes, the ac resistance between the electrodes varies in proportion to the degree of tilt.

Unfortunately the electrolytic sensors cannot be read using a dc voltage source. This would cause the deposits to form on the electrodes rendering them useless. Instead, an ac voltage of approximately 3 volts with a frequency of 1,000 Hz is fed to the sensor. The ac voltage from the center electrode is in proportion to the tilt of the sensor.

If one were to use the electrolytic tilt sensor, I can suggest one way to set up the information flow. Connect the ac output of the tilt sensor to a bridge rectifier to obtain a dc-equivalent voltage. The dc voltage is fed to a voltage-controlled oscillator (VCO). The VCO output frequency varies in proportion to the input voltage. The output of the VCO is transmitted over a radio link to a receiver on the motion platform. The receiver reads the frequency (tilt) and activates a proportional control to tilt the platform.

Spectron Inc. offers an integrated circuit, the SA40011, that simplifies interfacing electrolytic tile sensors. The dc output from the SA40011 can be fed to a VCO as described before.

Again implementing tilt to the operator is the difficult part of the system. Proportional pneumatic or hydraulic systems can be employed to the seat to provide tilt.

Greater video range

The video range of our small transmitter is approximately 300–2,000 feet. That's about the limit for a street-legal transmitter operating on TV broadcast bands. Obviously, for longer distances another system needs to be employed. It's called ATV.

Amateur Television (ATV) has been around for a number of years. It's been a method for radio amateurs to communicate via two-way television. ATV had been the province of the elite radio hobbyist due to the expensive cost of equipment. However, recent advances in solid-state technology have changed that.

The components for a 5-watt ATV system can be purchased for $200, excluding TV monitor and video camera. The video camera on Golem I and the LCD TV in the HMD are suitable for ATV use. Less powerful ATV systems, ¾ watt, can be purchased for under $100. That is still 7 times more powerful than the transmitter currently used on the Golem I.

A technician-class amateur license is required to operate these systems legally in the United States. Currently, the technician-class license no longer requires a knowledge of Morse code. Interested readers should contact a local amateur radio club for more information, or you may write:

American Radio Relay League (ARRL)
Newington, CT 06111
(203) 666-1541

A 5-watt ATV system can transmit a distance up to 30–40 miles, depending on local radio interference, terrain, weather, etc.

More models

With the experience gained in building this T-bot system, the reader can build other models. The company that makes the Erector Sets has revitalized and updated them and brought a number of interesting kits to the market. The kits are called Meccano-Erector

Sets that include motors, gears, and pulleys, along with the standard Erector Set materials.

There are standard kits for building trucks, cars, motorcycles, land movers, etc. The kits provide a good springboard for building exotic T-bot explorers, and kits are available locally through Toys R Us dealerships.

Parts list

☐ Microminiature B/W video camera—$85.00

☐ TV transmitter (kit)—$45.00

☐ Radio control system
  • 2-channel receiver
  • 2-channel transmitter
  • Xtals
  • 2 servo motors (42-oz torque)—$50.00

☐ Velcro material, 1-foot length—$4.50

Available from:
Images Comapny
POB 140742
Staten Island, NY 10314
(718) 698-8305

Mobile platforms

Platforms are the foundation for mobile robots. There are two options: build or buy. If one has good mechanical ablity or is willing to learn, building a platform from scratch offers distinct advantages. The platform is designed and built for the specific task and purpose of a specific robot. One has an almost unlimited choice of drive motors, gear boxes, mechanical linkage, power supplies, etc.

Buying a mobile platform relieves one from building a platform, but one is left with the gear ratio and power and speed designed for a different purpose. Here is a case in point: most electric cars move too fast. If one doesn't have mechanical ability, this is the way to go. Typically, one buys a radio-controlled (RC) electric car. The RCs are stripped from the unit, and the electrical connections (wires) to the steering control and drive motor are retained.

Here are some things to keep in mind when purchasing an electric car for conversion. First, don't choose a car that's too small or lies too close to the ground. The size will make it difficult to fit sensor systems and microcontrollers onto the chassis. If the car lies too close to the ground, it will get stuck easily. Choose a car with a high wheel base.

Figure 10.1 shows an electric car that is a bad choice for turning into a mobile robot. It's too small to carry substantial weight, and notice how low to the ground it lies. This car will get stuck easily. Figure 10.2 shows a better choice. The platform is larger, can fit more components, and has a high wheel base.

Stepper motors

If one wants to build a platform, stepper motors make excellent drive motors. Some of the advantages of a stepper motor are as follows. Because a stepper motor turns in precise increments per step, a microcontroller can calculate the distance traveled by counting the clock pulses given to the stepper motor and knowing the diameter of the drive wheel. If two stepper motors are used (one on each side of the robot, for locomotion and steering), precise turns are also possible.

Figure 10.3 is an electric-equivalent circuit of a unipolar stepper motor. The stepper motor has six wires coming out from the casing. We can see from Fig. 10.3 that three leads go to each half of the coil windings, and the coil windings are connected in pairs. If you just picked this stepper motor and didn't know anything about it, the simplest way to analyze it is to check the electrical resistance between the leads. By making a table of the wire colors and resistances measured between the leads, you'll quickly find which wires are connected to which coils.

■ **10.1** *Small electric RC car unsuitable for conversion*

■ **10.2** *Large electric RC car suitable for conversion*

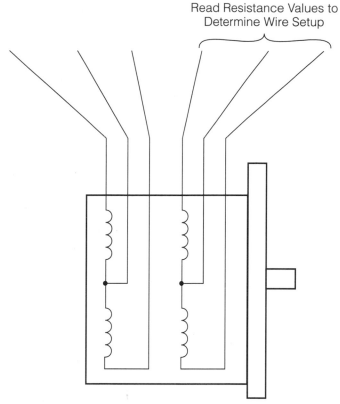

■ 10.3 *Schematic of six-wire unipolar stepper motor*

On the motor we are using, there is a 110-ohm resistance between the center tap wire and each end lead, and 220 ohms between the two end leads. A wire from each of the separate coils will show an infinitely high resistance (no connection) between them. Armed with this information, you can just about tackle any six-wire stepper motor you come across. The stepper motor we are using rotates 1.8 degrees per step.

UCN-5804

Figure 10.4 is a schematic pin-out of the UCN-5804. The IC is designed to control and drive a four-phase unipolar stepper motor, such as the one we are using. The IC has a continuous output rating of 1.25 amperes per phase at a maximum voltage of 35V. This is more than enough power to run our 24-volt stepper motor.

The UCN-5804 internal logic sequences its output pins in time with a square wave pulse delivered to pin 11. Each square wave

171

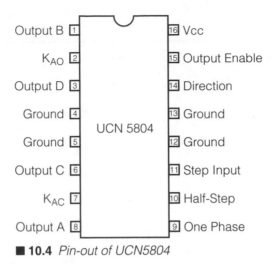

Output B ①   ⑯ Vcc

K$_{AO}$ ②   ⑮ Output Enable

Output D ③   ⑭ Direction

Ground ④   ⑬ Ground

UCN 5804

Ground ⑤   ⑫ Ground

Output C ⑥   ⑪ Step Input

K$_{AC}$ ⑦   ⑩ Half-Step

Output A ⑧   ⑨ One Phase

■ **10.4** *Pin-out of UCN5804*

pulse (High to low transition) delivered to this pin increments the stepper-motor sequence.

When you reach the end of the table, the sequence repeats, starting from the top of the table. To reverse the stepper-motor direction, start the sequence from the bottom, and work towards the top.

Pin 15 is the output enable. When this pin is held high, all outputs on the IC are disabled (off). If this function isn't required by your circuit or system, this pin should be tied to ground (low).

Pin 14 is the direction. When this pin is tied low or connected to ground, it will follow the sequence shown in either Table 10.1 or 10.2, starting from the top line and working downward. When this pin is tied high (+5V), it will reverse the sequence direction starting from the bottom and working its way to the top.

Using the UCN5804

Figure 10.5 is a schematic using the UCN5804. The clocking signal is provided by the 555 timer. The clocking signal may be increased or decreased using potentiometer V1. Varying the frequency of the clock signal directly controls the speed of the stepper motor. Chapter 7 showed how the BASIC Stamp microcontroller could also provide the clocking pulses to drive a stepper motor.

In this schematic, three manual on-off switchs control additional functions. The pins that the switches are connected to can also be controlled by the I/O pins off the BASIC Stamp microcontroller. The switch connected to pin 15 is the enable pin. When brought high, this pin disables the output of the UCN5804 chip, stopping the stepper motor.

The switch connected to pin 14 controls the shaft's direction, clockwise or counterclockwise. The switch connected to pin 10 is the step/half-step control. When pin 10 is brought high, the chip operates in the half-step mode.

This mode doubles the resolution of the stepper motor. For instance, the motor we are using rotates the shaft 1.8 degrees per step. When operating in the half-step mode, the shaft rotates .9 degrees per step. When operating in the half-step mode, the overall rotation speed (rpm) of the shaft will be one-half of the speed of the full-step mode.

Connecting a wheel to a stepper-motor shaft

Connecting a drive wheel to a shaft can become a major problem, and a simple solution is provided (see Fig. 10.6). Purchase a

■ Table 10.1

				Full-Step Sequence
A	B	C	D	Output pins on UCN5804 (see Fig. 10.1)
on	-	-	-	
-	on	-	-	
-	-	on	-	
-	-	-	on	

■ Table 10.2

				Half-Step Sequence
A	B	C	D	Output pins of UCN5804
on	-	-	-	
on	on	-	-	
-	on	-	-	
-	on	on	-	
-	-	on	-	
-	-	on	on	
-	-	-	on	
on	-	-	on	

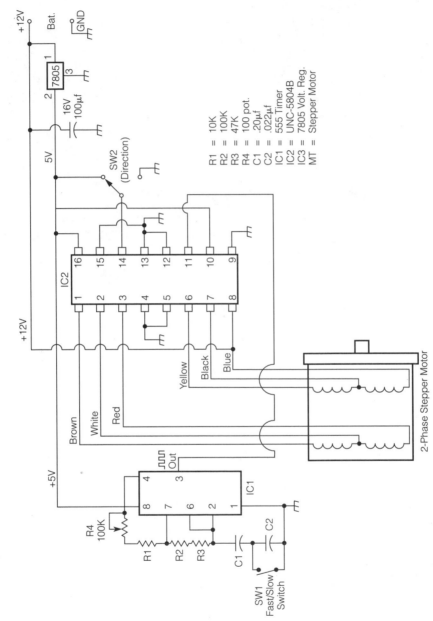

■ **10.5** *Schematic of basic stepper motor driver circuit*

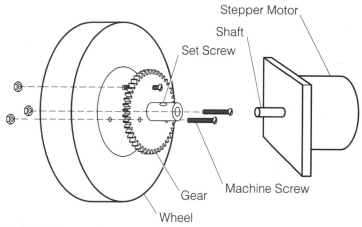

Stepper Motor

Shaft

Set Screw

Machine Screw

Gear

Wheel

■ **10.6** *Connecting wheel to motor shaft*

large-diameter plastic gear with a set screw. The mounting hole on the gear should match the shaft diameter of the stepper motor. Center the gear on the wheel. Drill three holes 120 degrees apart, through the gear and wheel. Mount the wheel to the gear using three machine screws, washer, and nuts. Next mount the wheel gear assembly to the stepper-motor shaft using the set screw.

Legos

Some budding robotists use Legos to prototype robotic platforms and designs (see Fig. 10.7).

Meccano kits

While not the snap-together functionality of Legos, Meccano kits (erector sets) can be used for prototyping also (see Fig. 10.8).

Capsela kits

This is another snap-together robotic system. I haven't played with the system, but it looks interesting (see Fig. 10.9).

Other kits

If you look around toy stores and hobby shops, you will find other kits that can provide foundations for robotic inspiration (see Fig. 10.10).

■ **10.7** *Lego kit*

■ **10.8** *Meccano kit (Erector set)*

■ **10.9** *Capsela kit*

■ **10.10** *Wood car kit*

Walker robots

Walkers are a new class of robots that imitate the locomotion of animals and insects. Walkers walk using legs. Locomotion by legs is hundreds of millions of years old. In contrast, wheels are a relatively new science, being about 7,000–10,000 years old. Wheels are good, but they require a relatively smooth surface to ride upon. Just look at the road and highway infrastructure of any developed country.

Why build walkers?

Walkers have the potential to transverse rough terrain that is unpassable by wheeled vehicles. It is with this in mind that robotists are developing walker robots.

179

Imitation of life

Sophisticated walkers imitate insects, crabs, and sometimes humans. Biped walkers are still rare, requiring a good deal of engineering science.

Six-legged tripod gate

Using a six-legged model, we can demonstrate the famous tripod gate used by the majority of legged creatures. In the following drawings, a dark circle means the foot is planted on the ground. A light circle means the foot is up and moveable.

Figure 11.1 shows our creature at rest. All feet are on the ground. The creature decides to move forward. To step forward, it lifts three of its legs (see Fig. 11.2), leaving its weight on the remaining three legs. Notice that the legs supporting the weight are in the shape of a tripod. This is a stable weight-supporting structure, and our creature is unlikely to fall over. The three lifted legs are free to move, and they move forward.

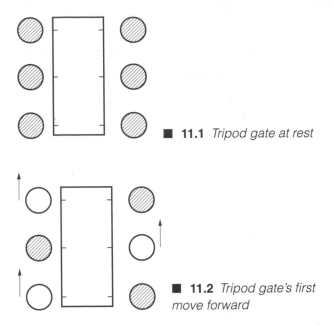

■ **11.1** *Tripod gate at rest*

■ **11.2** *Tripod gate's first move forward*

Figure 11.3 shows where the three lifted legs move. Now the creature's weight shifts from the stationary legs to the moveable legs (see Fig. 11.4). Notice that the creature's weight is always supported by that tripod position. Now the other legs move forward and the cycle repeats. This is called a *tripod gate* because the creature's weight is always supported by a tripod positioning of legs.

Creating a walker robot

There are a lot of toy walkers around (see Figs. 11.5 and 11.6). These walkers move their legs using a rotary cam mechanism. However, creating a walker robot that imitates a tripod gate requires a minimum of six servos and possibly twelve servos.

The need for so many servos is that each leg on the walker needs to have two axes (degrees) of freedom, one to move up or down and the second to move (swing) forward and back.

Two-servo walker

The walker robot we will make is a compromise. It's not a tripod gate walker, but it's still above a cam walker. The inspiration for this robot has come from Mark Tilden (see BEAM Robotics, Chapter 8). Mark has created this fabulous little walker using two tiny gearboxed motors and his nervous net technology. Our walker uses lightweight servo motors and a BASIC Stamp.

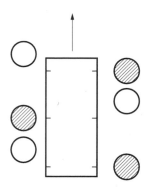

■ **11.3** *Tripod gate's second move, shifting weight*

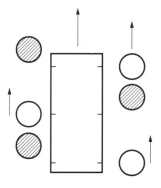

■ **11.4** *Tripod gate third move*

■ **11.5** *Toy walker uses cam to move legs*

■ **11.6** *Toy walker uses cam to move legs*

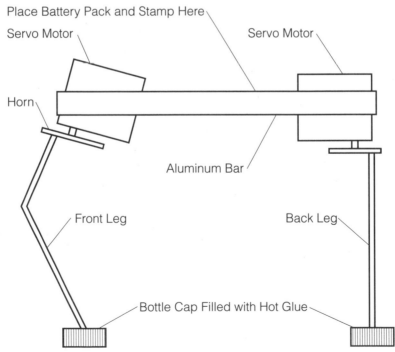

Place Battery Pack and Stamp Here

Servo Motor

Servo Motor

Horn

Aluminum Bar

Front Leg

Back Leg

Bottle Cap Filled with Hot Glue

■ **11.7** *Side view, two-servo walker*

Construction

For the frame, I used two aluminum bars $\frac{1}{2}$" wide, 6" long, and $\frac{1}{8}$" thick. The servo motors are hot glued to the aluminum bars (see Fig. 11.7), one aluminum bar on each side of the servo motors. The front servo motor is tilted upwards at about 30 degrees.

The legs for the robot are made from $\frac{1}{8}$" square steel rods 12" long. Figure 11.8 illustrates how to bend the legs. The front leg is bent backwards so that the weight of the servo motor is centered.

The legs are mounted on the servo using the servo horn. Two holes are drilled in the $\frac{1}{8}$" square steel to accept a 0–80 machine screw. The machine screw is passed through the servo horn and leg and is secured using a 0–80 machine nut.

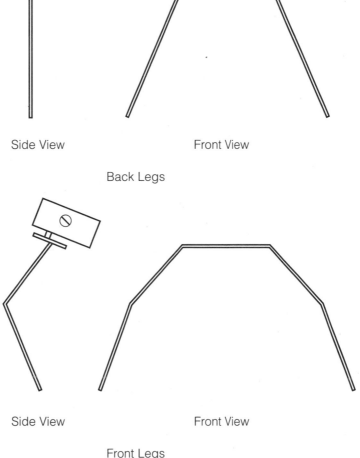

Side View   Front View

Back Legs

Side View   Front View

Front Legs

■ **11.8** *Front and back legs*

The bottom of the legs must have a wide area and be flat. When first building this project, I left the steel legs as they were. Unfortunately, the walker couldn't walk. The sharp edges of the legs would catch on something, tipping the entire robot over. After a number of false starts, I realized that the problem was not in the program or servo mounting but in the bottom of the legs. My solution was in finding four bottle caps, placing a bottle cap under each leg end, and filling the bottle cap with hot glue to secure the cap to the leg. As soon as I did this, the walker walked.

Electronics

Figure 11.9 shows the schematic for the servo motors and BASIC Stamp. Notice that the 6-volt battery pack is powering the BASIC Stamp and the servo motors. The battery pack is +6V using 4 AA

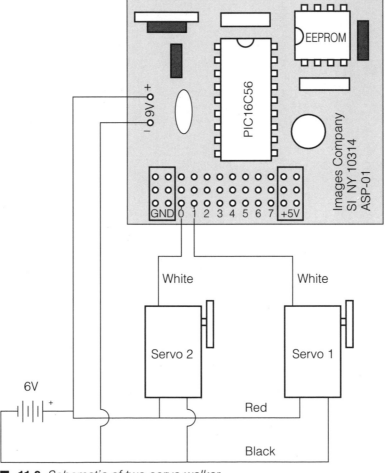

■ **11.9** *Schematic of two-servo walker*

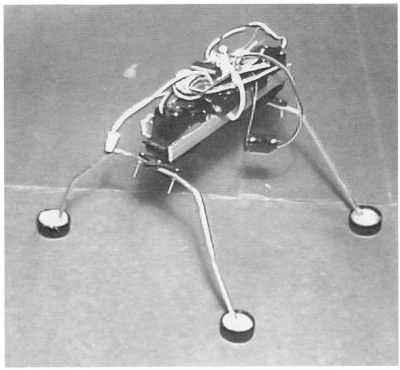

■ **11.10** *Two-servo walker robot complete*

batteries. The battery pack and BASIC Stamp are secured to the aluminum frame as shown in Fig. 11.7. Figure 11.10 shows the completed walker ready to run.

Stamp Program

The Stamp Program controls both servo motors. There are plenty of I/O lines and programming space to improve and add to this basic walker.

```
REM Servo Walker
dirs = %00001111
Start:
pulsout 0,150
pulsout 1,150
pause 20
l1:
b1 = 0
for b1 = 0 to 25
pulsout 0,160
pulsout 1,140
pause 20
next
l2:
b1 = 0
for b1 = 0 to 25
```

```
pulsout 0,140
pulsout 1,160
pause 20
next
goto 11
```

Parts list

☐ Servo motors

☐ BASIC Stamps

☐ Aluminum bars

☐ 1/8" steel rods

Available from:

Images Company
POB 140742
Staten Island, NY 10314
(718) 698-8305

Solar-ball robot

12

THE INSPIRATION FOR THIS ROBOT ORIGINALLY BEGAN with Richard Weait of North York, Toronto. Richard created a light-seeking robot in a transparent globe (ball). More recently, Dave Hrynkiw from Calgary, Canada, picked up the ball (so to speak) and developed a series of light-seeking mobile solar-ball robots.

There are two features to this mobile robot that are interesting (see Fig. 12.1). First is the method of locomotion. Inside the globe is a gearbox. One end of the gearbox's shaft is secured and locked to the inside of the inner surface of the transparent globe. The shaft being locked cannot rotate, which forces the gearbox itself to rotate. The gearbox is heavy, which moves the center of gravity of the sphere forward. In doing so, the sphere moves forward.

When at rest, the weight of the gearbox keeps it at bottom dead center (the gearbox facing down), and the ball resists rolling. When the gearbox is activated, the box begins to rotate inside the globe. This moves the center of gravity of the ball forward, causing the ball to roll forward.

The second feature relates to the power supply for the gearbox. The original solar robots had an onboard power supply that provided intermittent power to the gearbox. (For more information on this type of power supply, see Chapter 3.) The onboard power supply consists of a solar cell, main capacitor, and a slow oscillating or trigger circuit. When exposed to sunlight, the solar cell begins charging the circuit's main capacitor. When the capacitor reaches a certain voltage, a trigger circuit dumps the store electricity through a high-efficiency motor connected to the gearbox, causing the robot to move forward a little.

This solar-ball robot uses a similar gearbox assembly, but for power uses two standard AA batteries. The disadvantage to batteries is that they must be replaced when worn out. The advantage, however, is that they supply continuous power to the robot, allowing

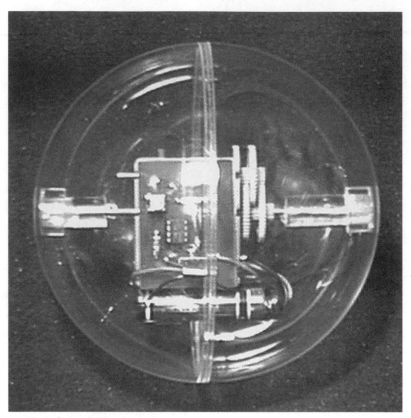

■ **12.1** *Solar ball robot*

one to easily study its behavior (mainly phototropism), locomotion, and mobility.

With the original solar-ball robot, one needs to use time-lapse photography to study these effects. The charging of the capacitor takes a few minutes, depending on the intensity of sunlight. When the electricity is discharged into the motor, the robot lurches forward a short distance. For example, 10 hours of motion with the original solar ball can be compressed into a few minutes of study with this robot.

While this particular robot doesn't incorporate the electronics for an onboard power supply, it still uses a light trigger. The circuit shown in Fig. 12.2 controls the power from the batteries to the gearbox motor. The circuit reads the level of illumination that the robot sees. If the light level is high enough, it turns on the motor to the gearbox. The trip level of the circuit is user adjustable using potentiometer V1.

Gearbox

Before we get into the construction of the robot, let's first look at the gearbox (see Fig.12.3). Physically, this gearbox is smaller than many gearboxes and is easier to fit inside the sphere. It has a 1000:1 gear ratio. The higher the gear ratio the slower the robot will move.

In the prototype, the gearbox is set to the 1000:1 ratio. You can use any gearbox that can fit and rotate inside the 5½" transparent sphere. Choose one with a high gear ratio that delivers a low rpm (7 rpm).

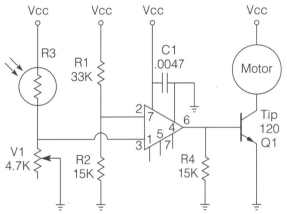

■ **12.2** *Schematic of solar ball circuit*

■ **12.3** *Gearbox 1000:1 ratio*

Robot construction

The shell is the first component for consideration. It must be transparent and large enough to hold the gearbox and electronics. The shell used in my prototype has a diameter of $5\frac{1}{2}$ inches. Snap-together transparent spheres are available in many hobby and craft stores. These hobbyists use them to enclose holiday ornaments. If you cannot find a suitable shell locally, you can purchase one from Images Company (see the supplier's list at the end of this chapter). The plastic shell is fragile. Do not have your robot try to climb or fall down stairs; it is sure to crack and break.

Separate the two halves of the shell. The first job is to locate the center of each half sphere. This is where the shafts of the gearbox will be connected. Locating the center at first appears much easier than it actually is. To find the center, I was forced to trace the diameter of the shell on white paper, then draw a box around the drawn circle that touched the circle on four sides (see Fig. 12.4). Drawing diagonal lines from the corners of the box, I was able to locate the circle center. The half sphere is then positioned over the drawn circle. If you hold your head directly over the half sphere, you may be able to eyeball the center and mark it on the sphere with a magic marker. I tried once or twice with less than ideal results. Finally, I taped the paper on a $\frac{1}{2}$" piece of

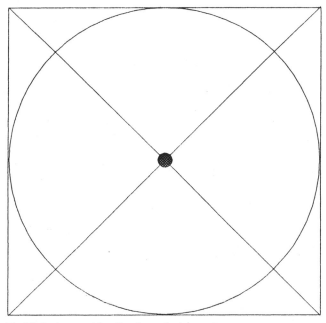

■ **12.4** *Layout for finding circle center*

wood and drilled a small hole at dead center. Then I placed a small dowel, about 2.5" long, in the hole, making sure it was perpendicular to the wood. Place the half sphere over the fixture, lining up its diameter with the drawn circle; the dowel locates the center of the sphere fairly accurately. Mark the center of one-half sphere, then the other.

The next step is to make a drive-locking fixture in the sphere that prevents the gearbox shaft from rotating freely inside. With the shaft locked, it forces the gearbox itself to rotate inside the sphere, changing the center of gravity and moving the robot along. The drive fixture must at the same time allow the sphere to be assembled or unassembled at will. The system I devised is illustrated in Figs. 12.5 and 12.6. Although I built all the drive components out of transparent plastic on the prototype, you can fabricate the parts out of other materials like brass and wood.

The first component is a small length of tubing $\frac{5}{8}$" Outside diameter (OD), 1/2" Inside diameter (ID), and about $\frac{3}{8}$" long.

This tubing is glued to the center of the half sphere, using the marks as a guide.

Inside the tubing, glue a $\frac{1}{2}$" diameter half round about $\frac{3}{8}$" long. This piece may be glued inside the tubing before the tubing itself is glued into the sphere.

Next, cut a small length of $\frac{1}{2}$"-diameter solid rod.

On one end of the rod, a $\frac{3}{8}$" half section is removed. This is accomplished using a hacksaw or coping saw. First make a cut directly down the center of the rod about $\frac{3}{8}$" deep. Then make a horizontal cut to remove the half section. Check to make sure this shaft fits easily into the $\frac{3}{8}$" tube and half round assembly inside the half sphere. If not, file the cut end until it does. At the opposite end of this rod, drill a hole down the center that will fit the shaft from the gearbox.

Note: on the prototype robot, I made the second shaft a drive connection also. Only when the robot was finished did I realize that this was unnecessary. A single drive connection works just as well as a double.

The second half sphere is easier to make. Glue a small length of $\frac{5}{8}$"-OD, $\frac{1}{2}$"-ID tubing to the center of the half sphere, using the mark as a guide. Cut a small length of $\frac{1}{2}$"-diameter solid rod. Check to make sure the shaft fits easily into the $\frac{5}{8}$" tubing. If not, obtain a small piece of 100-grit sandpaper. Wrap the sandpaper around $\frac{1}{2}$" length on the end of the shaft. Twist the sandpaper

191

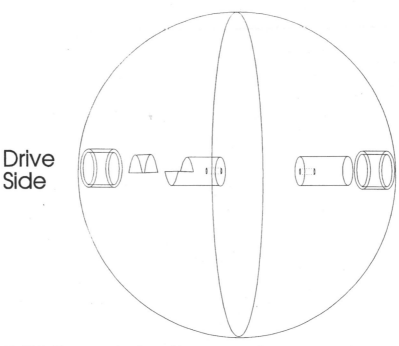

Drive Side

■ **12.5** *Transparent sphere drive components*

around on the shaft to sand the end. Continue sanding until the end of the shaft fits easily into and out of the tubing. Next, drill a hole down the center that will fit the shaft from the gearbox.

We want the gearbox to be positioned in the center of the sphere. Place the shaft of the gearbox in the plastic rod. Place the rod in the tubing on half the sphere of the globe.

Position the gearbox so that it will lie in the center. Mark the depth the gearbox shaft must go into the plastic rod on the gearbox shaft. Remove the gearbox shaft. Mix a small amount of 2-part epoxy glue. Coat the gearbox shaft with the epoxy glue and insert it into the plastic rod. Let the glue set before proceeding.

Once the glue has dried on the first shaft, we must glue the other plastic rod on the opposite side of the shaft. Position the glued rod into the half sphere. Place the other plastic rod on the opposite shaft. Place the other half sphere together with the first. Gauge the depth the gearbox shaft must be inserted in the plastic rod, then add another $\frac{1}{8}$" of depth for error. Glue and let set. Check your work while the glue is setting on the second shaft to ensure that you can close the sphere properly.

Electronics

The electronic circuit is a light-activated on-off switch. When the ambient light level is low (user adjustable), the circuit shuts off power to the gearbox. The user adjusts the sensitivity of the circuit using potentiometer V1.

There is nothing crucial about the circuit. If you do not wish to purchase or make the PC board, the circuit may be wired and assembled on a standard breadboard.

How it works

The circuit configures a CMOS op-amp as a voltage comparator. A comparator monitors two input voltages. One voltage is set up as a reference voltage called "Vref." The other voltage is the input voltage called "Vin," which is the voltage to be compared. When the Vin voltage falls above or below the Vref, the output of the comparator (pin 6) changes states.

The two input voltages are applied to pins 2 and 3. Pin 2 (inverting input) is connected to a reference voltage of approximately 1.5V, using a simple voltage divider made of resistors R1 and R2.

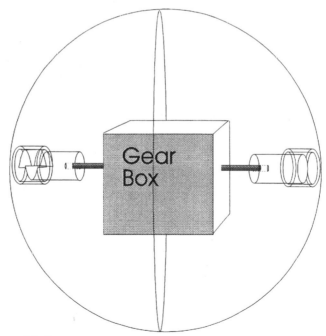

■ **12.6** *Gearbox placement in sphere*

Photosensitive resistor R3 makes up another voltage divider in conjunction with potentiometer V1, which is connected to the noninverting input (pin 3) of the op-amp.

There is no feedback resistor between the output (pin 6) and either of the inputs (pins 2 and 3). This forces the op-amp to operate at its open loop gain.

A CdS photoresistor is used as the light sensor. A photoresistor changes resistance in proportion to the intensity of the light that falls on its surface. The CdS produces its greatest resistance in total darkness. As the light intensity increases, its resistance decreases. In the circuit, the CdS cell is part of a voltage divider. The changing resistance of the CdS cell changes the voltage drop across the potentiometer V1, which is connected to pin 3. As the light intensity increases, the resistance of the CdS cell decreases, which increases the voltage drop across the potentiometer. This increased voltage drop is seen as a raising voltage. The trigger voltage can be set for different light levels using the potentiometer.

The electronic circuit is not crucial. You can construct the circuit using point-to-point soldering on a prototyping breadboard. A PCB board is available from a kit, or you can make it yourself. The PCB artwork is illustrated in Fig. 12.7. Parts placement on the board is shown in Fig. 12.8

Once the circuit is complete, you need to adjust the light level that will activate the circuit using potentiometer V1. Make temporary connections to the gearbox motor using alligator clip wires. Power to the circuit and gearbox is obtained from two AA cells, and the AA cell pack is glued to the back of the gearbox during final assembly. Make sure the battery pack has a battery clip for easily disconnecting and connecting power.

When making the light-level adjustment, use a low level of light to activate the robot. When the robot is on the floor, if the light level is set too high it will stop every time it passes under a shadow.

Putting it all together

Once the circuit is adjusted, you are ready for the final assembly. Glue the AA battery pack to the back of the gearbox making sure that no glue comes into contact with any of the gears. The electronic circuit board is glued to the front of the gearbox, again making sure that none of the glue touches any of the gears. Connect the power supply. At this point the gearbox will probably start turning. To load

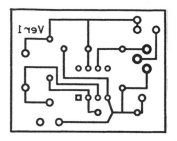

■ **12.7** *PCB layout*

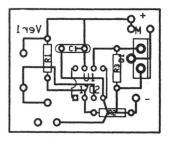

■ **12.8** *PCB parts placement*

the mechanism inside the robot, bring all the parts into a dark room to deactivate the circuit. Load the assembly inside the sphere.

Take the robot out into the light. The gearbox should become active. Place the robot on the floor. The robot should travel toward or in the direction of light. If the robot does the opposite, stop the robot, remove the gearbox and electronics, and reverse the wires leading to the motor.

Locomotion

I was pleasantly surprised when I began observing this robot. I originally thought it would become trapped easily. Not so. When the robot enters a corner and stops, the gearbox inside begins swinging all the way up and over, radically shifting its weight over top dead center and moving the robot out of the corner.

Advancing the design

When I originally designed this robot, I planned to use a steering mechanism to track a light source. However, the small steering mechanism didn't have enough weight to turn the robot in any direction quickly. In the long run, other factors (terrain, obstacles, etc.) affect its direction. Hence I removed the steering. But this is still a good research area for advancing the overall design.

Adding higher behavior module

As the robot stands, when a certain level of light is reached it becomes active. We can add a higher behavior mode, feeding, by adding a few more components (two solar cells and steering diodes) and another comparator circuit. The second comparator circuit will deactivate the motor when the light illumination level becomes high enough, allowing the solar cells to charge the AA batteries, which will be changed to nicads.

Figure 12.9 illustrates the behavior. When the light level is low, the robot is off, or we can say it is in a resting mode. As illumination increases, it reaches a point where the motor turns on and the robot enters its searching mode. When the light level increases significantly beyond this point (searching mode) the second comparator turns off power to the gearbox motor, allowing the two solar cells to charge the AA nicad batteries, which triggers the feeding mode.

If anyone plans to add this feeding behavior circuit, keep track of the current drain to the comparator circuits. It must not exceed the current supplied by the solar cells, or obviously no charging to the nicads will occur.

Parts list

☐ (1) 5½" transparent plastic globe (see earlier text in this chapter)

☐ (1) gearbox (see earlier text in this chapter)

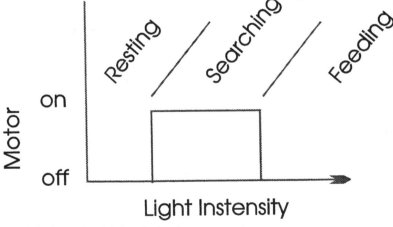

■ **12.9** *Layering higher behavior*

- ☐ (1) 6" length of $\frac{1}{2}$" solid plastic rod
- ☐ (1) 3" length of $\frac{5}{8}$"-OD $\frac{1}{2}$"-ID plastic tubing
- ☐ (1) 1" length $\frac{1}{2}$" half-round plastic rod

Electronics

- ☐ (1) 5V CMOS op-amp ALD 1702 or equivalent
- ☐ (1) 33K-ohm $\frac{1}{4}$-watt resistor
- ☐ (1) CdS photoresistor
- ☐ (1) 4.7K-ohm potentiometer (PC mount)
- ☐ (2) 15K-ohm resistors
- ☐ (1) .0047-uF capacitor
- ☐ (1) Tip120 NPN Darlington
- ☐ (1) PC board

A kit containing all of the above components is $45.00. Shipping and handling via UPS Ground Service is $7.50. New York state residents add 8.25% sales tax.

Available from:

Images Company
PO Box 140742
Staten Island, NY 10314
(718) 698-8305

All major credit cards accepted.

Underwater bots

13

Underwater robotics is an expansive field. Most underwater robots are designed for either salvage operations or exploration. In the future, robots may help farm the ocean. Not as well known but very interesting is that underwater robots may also be used as mock-ups to test space-faring robots. A neutrally buoyant robot is essentially weightless. Propellers and motors replace rockets, and inertia must be compensated. If you want to design a robot that will function in space, a good start is an underwater robot.

NASA has funded the development of Telepresence Remotely Operated Vehicles (TROVs) (see Fig. 13.1) and Autonomous Underwater Vehicles (AUVs). The TROV tests VR-based telerobotic techniques. Telepresence technologies are increasingly more important in exploration and hazardous duty, and telepresence technology will continue to grow in these fields and expand into others like entertainment.

Dolphins and tuna

Interestingly, studies are being conducted that examine the swimming motion and propulsion of fish. It is common knowledge that underwater animals move and swim more efficiently than a ship's propeller can move a ship. Want to prove this to yourself easily? Have you ever tapped on the glass of an aquarium filled with fish. The sudden noise sometimes causes the fish to dart around so quickly your eyes can't follow their movement. Imagine if you could design a ship that could move that fast, that suddenly. It's not surprising, then, that the U.S. government is funding some of these studies.

How efficient are fish at swimming compared to our current method of water propulsion? Let's glimpse at a partial analysis. In 1936 James Gray, a British zoologist, studied dolphins. His purpose was to calculate the power a dolphin needed to move itself at 20 knots, a speed at which dolphins are commonly reported to be able to swim. Gray's model of the dolphin was rigid, assuming that

■ **13.1** *NASA TROV Craft. Photo courtesy of NASA*

the water resistance for a moving dolphin is the same for a rigid model and flexible model. This is not true. Even so, accounting for this error, the conclusion Gray calculated is interesting. The dolphin is too weak, by a factor of seven, to attain the 20-knot speed. One may further deduce that the dolphin may be able to reduce its water resistance by a factor of seven to compensate. But this probably isn't the entire answer either, obviously.

Well, for the last 60 years no one has been able to prove or disprove Gray's calculations conclusively. Any swimming mechanism that mimics fishlike swimming is grossly inefficient. Recently, new studies are underway to again study fishlike swimming. With new computer technology behind this endeavor, scientists hope to answer these long-held questions.

Scientists at M.I.T. in Cambridge have been studying the bluefin tuna for the last four years. They created a 4-foot model "robot fish" that swims down the Ocean Engineering Test Tank Facility. The robot fish resembles the real fish. The skin is made of foam and lycra. The robot uses six external motors that are connected to pulleys and tendons within the robot. The fish moves and swims like a real bluefin tuna.

Swimming with foils

The tail of a fish is considered a hydrofoil. As the tail flaps from side to side, it pushes water backwards and propels the fish forward.

As the tail moves, vortexes are formed in the water behind it. It is believed that the vortex formation is key to understanding the greater efficiency of fish propulsion.

Dolphins are interesting. Their hydrofoil tail lies horizontally, so instead of moving its tail side to side as a fish, it moves its tail up and down. This creates the same efficient thrust in water, propelling the dolphin forward.

A penguin's method of swimming uses the thrust generated by its wings. Pictures of penguins swimming in water strongly resemble flying birds. There is a difference, though. With birds in flight, the beating of their wings must supply lift and forward thrust. The lift is necessary to counteract the force of gravity. With penguins, there is no necessity for lift. The density of water equals that of the penguin's body (neutral buoyancy), so the flapping of a penguin's wings simply needs to produce forward thrust.

Paddles and rows

Since we're looking at methods of locomotion in water, we might as well include paddles and rows. Ducks use their webbed feet as paddles swimming through water. Water beetles use their legs as oars and row themselves along like tiny boats.

What have we learned so far?

Studies at M.I.T. led researchers to use a fluid dynamic parameter known as the Strouhal number. For fish, the number is calculated by multiplying the frequency of the tail flapping back and forth, by the width of the vortex created in the water, divided by the fish's speed. A number of species of fish were studied. The results were that maximum efficiency is found when the Strouhal number lies between 0.25 and 0.35.

When the foils of the robot fish at M.I.T. were adjusted and reconfigured to generate a Strouhal number in this range, their efficiency jumped to higher than 86%. This is a major improvement compared to propellers that generate efficiencies around 40%.

Jumping in

There are two basic underwater robot projects. One involves modifying a toy submarine, the other building a robotic fish from scratch. First the submarine.

Submarine

There are a number of companies that make and sell hobby model submarines. Depending on the degree of sophistication of the model, it usually is radio controlled (RC) and is capable of submerging and surfacing (see Fig. 13.2).

In modifying a toy submarine, forget about RC and jump to wire control using an umbilical cord to the submarine. The umbilical cord can carry power and command and control signals.

These hobby submarines can be modified to create small telepresence systems. Primary modification is the addition of a color video camera. Most of these hobby submarines have open compartments where electronic gear can be stored (see Fig. 13.3).

Many of the systems used in the telerobotic car built in Chapter 9 can be implemented here. The one difference is the use of wire control instead of RC control.

Because these are "toy" subs, you will probably not let them loose in open water. The tiny propulsion motors in these submarines need calm water to function. Of course, these could be a starting point for more robust systems.

Are there any applications of these toy submarines beyond the experience gained from building and using basic underwater telepresence systems? I could imagine 10 or more toy telepresence submarines released in a swimming pool, each sub being remotely controlled by one person. I'm sure a number of underwater or space scenarios could be created for game play.

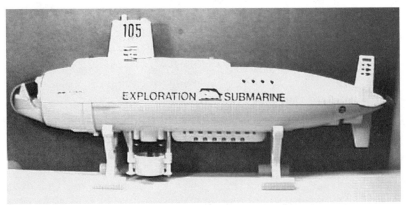

■ **13.2** *Toy submarine ready for conversion to TROV*

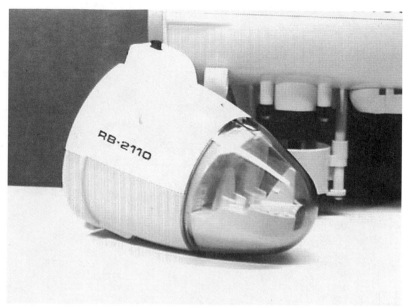

■ **13.3** *Open compartment for housing electronics*

Swimming by use of a tail

As stated earlier, most mechanisms that mimic fishlike movement are grossly inefficient. This model is no exception. However, information gleamed from resources like M.I.T. can be incorporated into the model (not done here) to improve overall efficiency. And if one plans on building like animal androids, this is as good as a place to start as any.

Rotary solenoid

The robotic fish pivots on a rotary solenoid (see Fig. 13.4). When activated, the solenoid rotates its top plate about 30 degrees. A spring returns the plate to its original position when deactivated.

The solenoid's top plate has at least two $\frac{3}{48}$" threaded holes that may be used for mounting objects to it. The bottom of the solenoid has two protruding $\frac{3}{48}$" threaded rods that can be used for mounting the solenoid. The solenoid is not as powerful as I would like, but it is strong enough to provide underwater propulsion.

Electronics

The electronic circuit uses a unijunction transistor Q1 (UJT 2646) to generate a slow stream of pulses (see Fig. 13.5). Timing of the pulses is determined by C1 and R1. The pulses pass through R4 to the base of Q2. Q2 is an NPN transistor 2N2222. The purpose of Q2 is to invert the pulse signal for input to IC1 pin 2. IC1 is a 555 timer

■ **13.4** *Rotary solenoid*

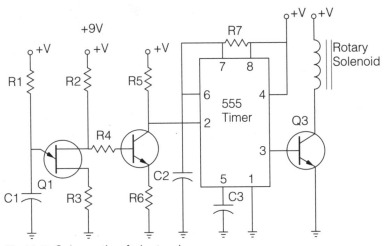

■ **13.5** *Schematic of electronics*

configured in monostable mode. IC1 shapes the pulse width, and the output of the 555 timer switches Q3 on and off. Q3 controls the current to the rotary solenoid that is used in the robotic fish.

The circuit is powered by a single 9-volt battery and is simple enough to hardwire on a prototyping-style PC board. Test the circuit by connecting it to the rotary solenoid before continuing. The time period of the pulse should be approximately one second.

Mechanics

To keep weight and mass down, most of the components are made out of aluminum. The first mechanism I used to convert the solenoid movement to a flapping fish tail is shown in Fig. 13.6. This turned out to be more complex than was necessary. Figure 13.7 shows the final tail assembly setup.

A $\frac{1}{8}$"-thick × $\frac{1}{2}$"-wide-by-$5\frac{1}{2}$"-long piece of aluminum bar is secured to the top plate of the rotary solenoid using two $\frac{3}{48}$ × $\frac{1}{4}$" screws. First drill two holes in the aluminum bar to match the holes in the top plate. Next, one hex nut is screwed flush to the underside of each screw head to prevent the screws from being driven too far down. If the screws are driven too far down, they will prevent the top plate from turning easily. Secure the aluminum bar to the top plate using the screws.

Fins are made by cutting a square of $1\frac{1}{4}$" aluminum diagonally. The fins to the tail are secured to the $\frac{1}{2}$" aluminum bar using a generous amount of hot glue. You may want to rough up the aluminum surfaces with sandpaper for a better bite before gluing.

The solenoid itself is secured to the end of a piece of aluminum $\frac{1}{8}$" thick by $1\frac{1}{4}$" wide by 2" long, using the two bottom $\frac{3}{48}$" threaded studs and a few $\frac{3}{48}$" hex nuts. On the front of the aluminum, the circuit and battery are secured (see Fig. 13.8).

Getting wet

Obviously, we have an exposed circuit and solenoid. To prevent water from damaging any components, cover the components using

■ **13.6** *Original tail assembly*

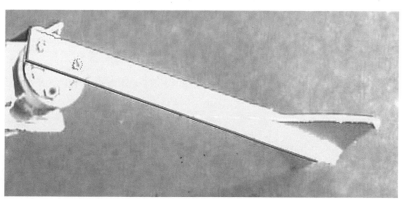

■ **13.7** *Final tail assembly*

■ **13.8** *Finished robot fish*

a thin, transparent plastic sandwich bag. The bag is secured to the tail using wire, but the bag should be such that the tail section can still move back and forth easily.

Before you dump the robot into the water, it should be made neutrally buoyant. If you dumped it in as is, the front-heavy robot would nosedive to the bottom of the water tank with the tail swishing back and forth ineffectively. Secure strips of styrofoam to the front of the model on the outside of the transparent bag using rubber bands. Place the model in water to test. When the model submerges and floats underwater level or almost level, you're ready to go. Turn on the power to the circuit and let the robot go.

Efficiency

This particular robot doesn't move with the grace or efficiency of a real fish, but it does move. I think efficiency can be improved by cutting the $\frac{1}{2}$" × $5\frac{1}{2}$" aluminum tail bar in half, then securing the halves back together using 2" of spring. This spring will allow the tail section to bend and flex and should create better thrust.

The robotic android fish

What separates a robot from an android is all in appearance. A robot looks like a robot and an android looks human or like the thing it was made to simulate. So an android fish should look like a fish.

Well, creating an android fish may not be as difficult as it may first appear. And the reason for this is that reasonable-quality fish coverings are available (see Fig. 13.9). These coverings are sold at magic and hobby shops, and these rubber fish can be cut open and the robotic mechanism inserted inside.

Some models are more realistic than others. I found one model made out of thick, soft rubber. While the appearance and texture of this model is excellent, the internal robotic structure to move the tail must be more powerful. A second fish covering that is less realistic but much thinner and therefore easier to move is a better choice.

Learn more about it

To learn more about fish-based propulsion systems, try reading the following sources: *Scientific American*: March 1995, "An Efficient Swimming Machine" by Micheal S. Triantafyllou and George S. Triantafyllou, and *Exploring Biomechanics* by R. McNeill Alexander, published by Scientific American Library, ISBN #0-7167-5035-X.

Components

R1	33K	
R2 & R6	100 ohm	
R3	470 ohm	
R5	10K	
R7	15K	
Q2	NPN 2N2222 transistor	
Q3	TIP120 NPN Darlington	
IC1	555 Timer	
C1 & C2	22-uF capacitor	
C3	.01-uF capacitor	
All the above components		$5.00
Rotary solenoid (with 4 nuts and 2 screws)		$5.95
Q1 2N2646 UJT		$5.95
⅛"-thk ½"-wide × 6"-lg aluminum bar		$1.50
⅛"-thk 1¼"-wide × 2"-lg aluminum bar		$1.00
All the above components		$19.95
Thin flexible fish covering		$6.50
Shipping and handling via ground UPS		$7.50

■ **13.9** *Rubber fish covering for robot fish*

Available from

Images Company
POB 140742
Staten Island, NY 10314
(718) 698-8305

All major credit cards accepted.

Aerobots

14

Aerobots (aerial robots) are a class of robots that can fly. They include lighter-than-air aircraft (blimps), helicopters, and airplanes. Some applications for aerobotics are autonomous flight, drones, warfare, surveillance, advertising, and telepresence.

Autonomous aircraft have a long history, the first ones being built in the early 1920s. One UAV (unmanned aerial vehicle) code named the "bug" was designed for warfare. The bug was about 12 feet long with a wingspan of 15 feet. Sophisticated flight control systems (for its time) included gyroscope, altimeter, with electric and pneumatic controls. The flight control system flew the craft 30–40 miles into enemy territory. When the desired distance was reached, the craft would jettison its wings, forcing the nose-heavy fuselage to fall to earth carrying a payload of 200 lbs of explosives. Unfortunately, WWI ended before the bug could see any action.

From this beginning, UAV's have been under continual development and refinement. The latest UAV saw action in the Persian Gulf war. Although the UAVs received little to no press, they flew over 300 sorties. They perform reconnaissance, damage assessment, and follow enemy weapons deployment. The most recognized and most sophisticated autonomous aircraft is epitomized by the self-guided cruise missile carrying nuclear warheads.

Telepresence flight control systems also have a long history, but not as long as UAVs. In the Second World War, the United States used remotely piloted aircraft to fly kamikaze missions. The old-style remote control systems are nowhere near the technical sophistication of today's systems. The old remote controls were unreliable, and the pilot needed to keep a visual eye on the remote aircraft to fly it accurately.

Today, remotely piloted aircraft have video cameras transmitting pictures back to the pilot, who may be situated anywhere in the world. The systems have developed into telepresence virtual reality systems.

The aerobot we will build is a flying telepresence blimp. The reason I choose this mode of flight (a lighter-than-air framework) over a model helicopter or model airplane is safety, silence, low cost, and ease of use.

Blimps are quiet, slow, graceful, and forgiving in flying errors. Safety was the major factor in my decision. If a blimp bumps into a person or object, there is little or no harm. The propellers on airplanes and helicopters, on the other hand, are potentially lethal weapons when in close proximity to human life.

The blimp we will build is limited to indoor use, and care must be taken in choosing components that are extremely lightweight. The payload capacity (lift) of the blimp is approximately 6 ounces. This means our RC receiver, propulsion, power supply, CCD camera, and video transmitter together must weigh in at or under this 6-oz weight restriction. Tough, but not impossible.

Lighter-than-air aircraft background

Lighter-than-air aircraft falls into three categories: rigid, semirigid, and nonrigid. Rigid aircraft have internal frames that are usually made from lightweight aluminum. The most famous of these crafts are the Zeppelins.

Semirigid aircraft have a rigid lower keel section, and a nonrigid envelope that is filled with helium is secured above it. Nonrigid aircraft are the ones we are most familiar with today. These are blimps. One of the more famous blimps is the Goodyear blimp used for advertising. Nonrigids are made of huge gas envelopes that take shape when the form is filled with helium gas.

Blimp systems

The most widely known uses for blimps today are in major football games, where they provide a bird's eye view of the game. Another popular use people are familiar with is advertisements.

While blimps may seem like old technology, scientists and engineers are still developing uses. For instance, the U.S. Army has used an unmanned airship called SASS LITE (Small Airship Surveillance System, Low Intensity Target Exploitation). The SASS LITE is used for border patrols. Recently the manufacturer stated that this 90-foot airship will be available for commercial ventures.

Helium balloons are capable of reaching the upper stratosphere, and one company has proposed building an air station 100,000 feet above the earth. The station would provide telecommunications links just like a satellite. However, the air station would cost 50% less than a similarly equipped satellite.

Robotic systems and telepresence systems have been put on model blimps for a number of years. We will review two ventures, one from the Robot Group and the other from Berkley's WEB Blimp shortly. What we will focus on accomplishing is placing a simple telepresence system on a model blimp. In reality, the telepresence system is a wireless, fly weight, portable surveillance system. Sensor feedback systems that could relay a sense of touch, for a "real" telepresence are not developed. Our simple system transmits video and sound, and the user or operator can move (fly) the blimp via RC.

The Robot Group—Austin, Texas

Robotic systems have been place on model blimps. The Robot Group, based in Austin, Texas, exhibited a robotic blimp at Robofest 1 in the fall of 1989. I'm sure that robotic systems have been placed on blimps before this for military and scientific purposes. However, the Robot Group represents private (nongovernment funded) exploration in this area. The Robot Group continues to develop and improve on the robotic blimp. In 1991 the computer blimp project called the Mark III used ultrasonic sensors and a neural network navigation system. Although the system fell short of design expectations, it did function properly. The Robot Group has a Web site on the Internet, where you can visit them to get the latest information. See Internet access at the end of this chapter.

WEB Blimp, Berkeley University

Space browsers is the name given to telepresence blimp systems being designed and built at Berkeley University, Department of Electrical Engineering and Computer Science. The blimps are used as avatars, or, as I prefer to call them, golems. The Berkeley group is striving for tele-embodiment systems. A true tele-embodiment system would require a complex sensor feedback system from the blimp avatar to the user. Currently the feedback system provides video and sound, and the user can maneuver the blimp via RC.

The most interesting aspect of this blimp is that it can be controlled over the Internet, hence the name WEB Blimp. The video is fed to the Internet via a video frame grabber with a CU-SeeMe format output. The WEB Blimp is made available through Berkeley's Web site (see Internet access at the end of this chapter).

Designing telepresence blimps as avatars and golems

Almost as good as being there! Robotic blimps or a reasonable facsimile have a good future in the telepresence industry. Suppose you wanted to look at some paintings in the Louvre in Paris, visit the Museum of Natural History in New York City, then jump over to the Smithsonian in Washington, DC, and finally check out the penguins on the Galodos Islands. And let's say you wanted to do all of this in a couple of hours.

One way this may be accomplished in real time is through the use of telepresence systems. One day in the future there will be sightseeing telerobots that you may jump into through a phone (or satellite) link and your home computer VR system. These robots will be located at many points of interest throughout the world.

The telerobots are not restricted to earth. There will be telerobots in space, underwater, and flying through the air. The Jason project is one underwater science adventure for schools. Through a satellite link, schools set up a communication link to scientists on a remote vessel. Students are able to learn what the scientists are doing, ask questions, and sometimes operate a TROV (telepresence remotely operated vehicle) via the satellite link.

To the moon

Lunacorp in Arlington, Virginia, has plans to place a civilian rover on the moon (see Fig. 14.1). For part of the time, the rover will be used as a telepresence system for earth-bound drivers (see Fig. 14.2). Unfortunately, operation cost is expensive, approximately $7,000 per hour. I don't know about you, but I'll plunk down $120.00 to drive a telepresence rover across the lunar sur-

Photo courtesy of LUNA Corp

■ **14.1** *LUNA Corp rover*

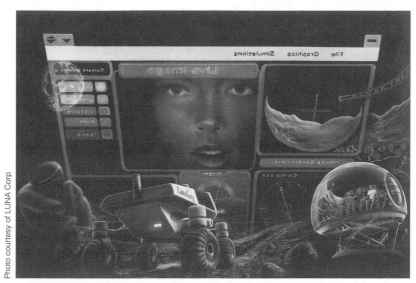

Photo courtesy of LUNA Corp

■ **14.2** *Artist's conception of LUNA telepresence system*

face for a minute. Lunacorp plans to get the rover to the moon in 2000: the site—Tranquility Base. But we are digressing away from our main topic of blimps.

Blimp parameters

Blimps need to meet certain design criteria to be used effectively for earth-bound terrestrial telepresence systems. The blimps must be completely safe around human beings and should be able to move through the same passageways used by people. The CCD camera transmitting video should be positioned at approximately eye level. It should be able to navigate through gentle crosswinds without difficulty.

A ballast system needs to be created that would allow the blimp to achieve neutral buoyancy on or through a number of floors in a building. If a ballast system is too difficult to implement, neutrally buoyant robot blimps can be positioned on each floor. Operators would simply switch to an unused telepresence robot held on a requested floor when wanting to change floors.

Because of the low weight of blimps, they have physical restrictions. For instance, the blimp would not be able to push open a door. Buildings would need to be modified to open doors and operate elevators electronically using remote control signals emitted from the blimp.

The blimp kit

The blimp we will construct is made of a tough mylar material that can be heat sealed using a household iron. There are many different styles of blimps one may build: a flying saucer-shaped, delta wing glider-shaped, or a typical "Goodyear" blimp. The one I recommend is the simplest of them all, a pillow-shaped blimp.

Making a pillow-shaped blimp is easy. Fold the sheet of mylar material in half (shiny side out). Heat seal the three open sides closed, leaving a little space that isn't heat sealed for a fill tube at the bottom, and you're finished.

Helium

Helium is sold in canisters from many party stores to make balloons. The canisters resemble those used to hold propane gas. If you don't have a local party goods store, look under helium or gas in the yellow pages to find a supplier.

Helium versus hydrogen

When I first began this project I thought about using hydrogen instead of helium, reasoning that hydrogen weighs about half of what helium weighs, so I could increase my lift by a factor of two, right? Wrong! I was correct in my assumption that hydrogen weighs about half of what helium weighs.

	English	Metric
Weight of hydrogen	0058 lbs/cubic foot	.09 kg/cubic meter
Weight of helium	.0110 lbs/cubic foot	.178 kg/cubic meter

I was incorrect in calculating the lift. Here's why. Lift is generated by the amount of air displaced by the helium (or hydrogen), just like an air bubble in water. Let's use the air bubble in water analogy. The air is less dense than the surrounding water, so the air bubble rises to the surface. Likewise, helium is less dense than the surrounding air, therefore it rises also. Think of the rising helium or hydrogen as floating on top of a much denser gas that we call air.

	English	Metric
Weight of air	.0807 lbs/cubic foot	1.29 kg/cubic meter

So what's the lift of a helium balloon with 5-cubic feet of displacement?

The weight of the displaced air equals 5(.0807) = .4035 lbs

The weight of 5 cubic feet of helium 5(.0110) = .0550 lbs

Lift = (.4035 lbs − .0550 lbs) ——> = .3485 lbs

That's quite a bit of lift! The reason is that we didn't subtract any weight for the balloon. If the balloon weighs .25 lbs, the usable lift (.3485 lbs − .25 lbs) is reduced to .0985 lbs or 1.57 ounces.

How does this compare to the lift using hydrogen? Well, the weight of the air displaced is the same.

The weight of 5 cubic feet of hydrogen 5(.0058) = .029 lbs

Lift = (.4035 lbs − .029 lbs) ———> = .3745 lbs

The difference in lift for a 5-cubic-foot balloon is:

.3745 lbs − .3485 lbs = .026 lbs or about ½ ounce

Because the difference in lift is small, it is not worth the added risk of using hydrogen gas! I recommend using helium gas only.

Size

The piece of mylar used to make a balloon after it's folded in half, lying down flat, measures 34″ × 56″. The weight of the material is 3 ounces (.1875 lbs). It's difficult to estimate how much helium the balloon will hold. To make a rough estimate, I use the volume of a cylinder. I know a pillow shape is not a cylinder, but, like I said, it's a rough estimate. First find the diameter. The material is 34 inches times two equaling 68 inches for the circumference. The circumference of a circle is two times pi (3.14) times the radius. If you do the math, the radius works out to 11 inches. The volume of a cylinder equals pi times the radius squared times the height. The height in this case is 56 inches. If you do the math, the volume equals about 12 cubic feet. The balloon will not be filled to its maximum capacity. In this case I'd estimate the balloon will hold about 70% of the calculated volume or about 8.4 cubic feet of helium gas.

Calculated lift:

Weight of air 8.4(.0807 lb/cubic ft) = .678 lbs

Weight of helium 8.4(.0110 lb/cubic ft) = .0924 lbs

Weight of Mylar Material 3 ounces or .1875 lbs

Lift —— > +.678 −.0924 −.1875 = .398 lbs or 6.37 ounces.

Construction

Construction is simple and straightforward. Essential to the construction of the blimp is being able to make a good heat seal. Cut a small section of mylar material from the large sheet to practice on. Fold the small section of mylar, shiny sides to the outside, dull sides together. Set the iron to a medium heat setting. Every time you adjust the temperature of the iron, allow at least 5 minutes for the iron to stabilize to the new temperature. If the temperature setting is too hot, the mylar material will melt and create holes. If the temperature setting is too cold, the heat seal will pull apart too easily. A good heat seal will not pull apart easily.

Allow the mylar material to cool for a minute before you test the heat-sealed seam. Keep adjusting and testing the heat setting of the iron on small mylar practice scraps until you find the right temperature. Once you found the right temperature setting, write it down for future reference.

We will make a pillow-shaped blimp. Fold the sheet of mylar material in half (shiny side out). Heat seal the three open sides closed, leaving a little space that isn't heat sealed for a fill tube at the bottom, and you're finished. The heat-sealed seam should be $\frac{1}{2}$" to 1" wide.

CCD camera

The CCD camera provides the video from the blimp (see Fig. 14.3). Naturally, weight is a consideration. This camera weighs a little over half an ounce. The overall size is $1\frac{1}{4}$" $\times$ $1\frac{1}{4}$" $\times$ $1\frac{1}{8}$". Light sensitivity is .03 LUX. Resolution is 430 TV lines. Output video is a standard NTSC (1V pp) signal. The camera can be powered from 9 Vdc to 12 Vdc maximum.

The current draw from the camera is approximately 100 mA, and a 9-volt transistor battery can power the camera. The battery weight (1.5 oz) is three times greater that the camera itself.

TV transmitter

There are a number of TV transmitter kits available, but there are two basic classes of transmitters. One type transmits the video and audio on one of the standard TV channels, and the TV's tuner picks up the signal and displays it. These transmitters have a limited range of a few hundred feet.

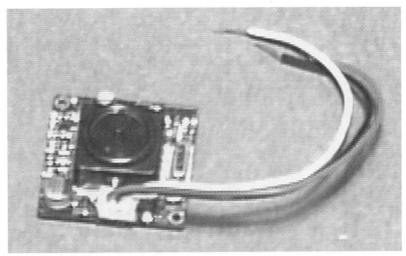

■ **14.3** *Lightweight CCD camera for telepresence system*

The second type is more expensive. This type transmits well above TV frequencies into the 900-MHz range, and the TV requires a down converter to display the video. The down converter receives the 900-MHz signal and down converts it to a standard TV frequency. These units have a much greater range and better fidelity. The unit used in this prototype transmits directly to a TV set on its VHF (Channel 14) channels (see Fig. 14.4).

Radio control (RC)

The radio control (RC) system is specially designed for blimps (see Fig. 14.5). It is extremely lightweight. The propulsion unit is a twin turbo fan that attaches to the underside of the blimp. Each turbo fan is bidirectional, and each is controlled by its own channel on the two-channel transmitter.

This helps the maneuverability of the blimp. While one turbo fan pushes forward, the other pushes backward, helping turn the blimp quickly. The pillow blimp ready for the telepresence system is shown in Fig. 14.6. Figure 14.7 is a closeup of the turbo fan, miniature CCD camera, and TV transmitter.

Going further

The blimp as it stands is a telepresence system. By placing autonomous navigation, we can convert the blimp into a flying robot.

■ **14.4** *TV Transmitter circuit*

■ **14.5** *Lightweight RC control system for blimp*

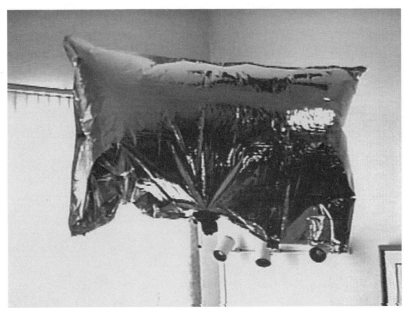

■ **14.6** *Pillow blimp*

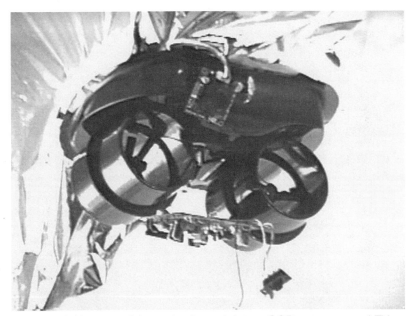

■ **14.7** *A closeup of the turbo fan, miniature CCD camera, and TV transmitter.*

Parts list

☐ Blimp kit—$19.95
☐ R/C control—$16.95
☐ B/W CCD camera—$90.00
☐ TV transmitter—$50.00

Parts are available from:

Images Company
POB 140742
Staten Island, NY 10314
(718) 698-8305

Internet access

RobotGroup Austin Texas—Neural Net Blimp
http://www.robotgroup.org/projects/mark4.html

WEB-Controlled Blimp at Berkley
http://vive.cs.berkeley.edu/blimp/

Web Blimp
http://register.cnet.com/content/features/quick/weblimp
http://utopia.minitel.fr/mpj/airships/">Marv's Airship Server

University of Virgina—Solar-powered airship
http://minerva.acc.Virginia.edu:80/secap/

U.S. competitor in Australia Solar Challenge
http://www.mane.virginia.edu/airship.htm

Intelligent Surveillance Blimp at the University of Virginia
http://watt.seas.virginia.edu/jap6y/isb/

Japanese Project—Solar-powered airship
http://www.aist.go.jp/mel/mainlab/joho/joh04e.html

Robotic arm

15

THE MOST WIDELY USED INDUSTRIAL ROBOT IS THE ROBOTIC arm. Robotic arms have many applications in manufacturing because the end manipulator of a robotic arm can be changed to fit particular industries. For instance, welding manipulators are used as spot-welding robots in manufacturing (automotive), spray nozzles for spray painting parts and assemblies (numerous industries), grippers for pick and place (electronics industry), to name a few.

As you can see, robotic arms are useful. Building a robotic arm from scratch is a difficult task, especially if you haven't physically seen a working robotic arm close up. It is easier to assemble a robotic arm from a kit, then interface the robotic arm to a host computer. In this case the host computer will be an IBM or compatible PC computer. The robotic arm is an OWI kit available from a number of electronic distributors. We will interface the robotic arm to the computer's printer (parallel) port.

The robotic arm (see Fig. 15.1) can move freely in two axes of motion. It can move vertically (up or down) and rotate clock-wise or counterclockwise. The gripper portion of the robotic arm can grasp and release small objects, but the arm is missing one important function—a rotating wrist.

This robotic arm is a simpler version of the Armatron robotic arm sold by Radio Shack (see Fig. 15.2). The Radio Shack arm has more features. For instance, the Radio Shack robotic arm has two speeds and has wrist rotation (clockwise or counterclockwise). Unfortunately, the Radio Shack arm achieves all of its motion with just a single motor and an ingenious concoction of gears and controls. The gearing controls make interfacing the Radio Shack robotic arm too complex.

Instead, the OWI robotic arm kit outlined here uses three small dc motors. The motors achieve a "wire" control robotic arm, wire control meaning that each robotic function (dc motor) is controlled by a wire (electrical power). Each dc motor controls a robotic arm function. The wire control makes it possible to build a controller unit for the arm that will respond to electrical signals. This simplifies the task of interfacing the robotic arm to a PC computer printer port.

Once interfaced to the printer port, the robotic arm may be operated in real time using a joystick or keyboard. In addition, the arm can be programmed in BASIC or QBASIC to perform a sequence of

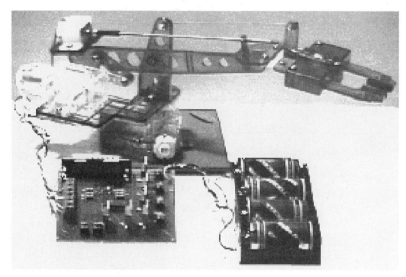

■ **15.1** *Robotic arm with PC interface*

■ **15.2** *Radio Shack Super Armatron robotic arm*

motions. The motion sequence may be repeated indefinitely as is the case in many animatronics.

The arm is made from lightweight plastic. Most of the stress-bearing parts are also made of plastic. The three dc motors used in the arm are small, low-torque motors. To increase the motor's torque, two gearboxes are built into the robotic arm. This allows motors to move the arm vertically and rotationally. Even with the gearboxes, the robotic arm is not capable of lifting or moving much weight.

The purpose of this project is to provide an opportunity for the experimenter to learn the mechanical structure of a simple robotic arm, seeing it operate, and learning how to program it via the computer. The experience obtained from this project can help someone build more robust robotic arms with greater capabilities from scratch.

Robotic arm assembly

The robotic arm kit has been thoughtfully laid out. The components are packaged in individual transparent plastic bags that are identified with an alphanumeric number (see Fig. 15.3). By carefully following the directions enclosed with the robotic arm, the construction proceeds smoothly (see Fig. 15.4). A few hours later, you have an operational robotic arm (see Fig. 15.5).

Test IBM interface

The stock robotic arm comes with a single 3-volt power supply. The first step in interfacing the arm to the computer is to replace the standard 3-volt power supply with a bipolar +/–3-volt power supply. The bipolar power supply may be from two 3-volt D-cell battery holders (see schematic, Fig. 15.6).

Each one of the three dc motors on the robotic arm will be controlled by two NPN Darlington transistors. Each NPN Darlington transistor controls the current flow to the motor, but in opposite directions. Figure 15.6 is a test circuit that you may construct before building the robotic arm interface. If neither transistor is turned on, the motor is off. Only one transistor (per motor) should be turned on at a time. If two transistors to the same motor are accidentally turned on at the same time, it will be the equivalent of creating a short circuit.

Use the test BASIC program to run a single motor (see BASIC listing dc Motor Control Test at the end of this chapter).

■ **15.3** *Packaged parts from OWI robotic arm kit*

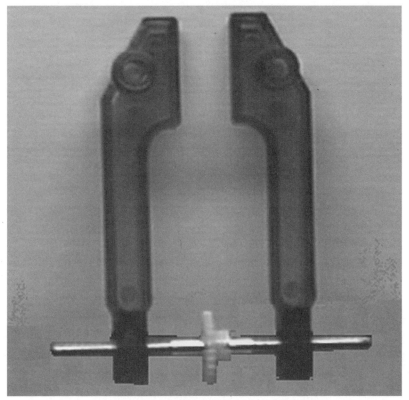

■ **15.4** *Gripper portion of robotic arm being assembled*

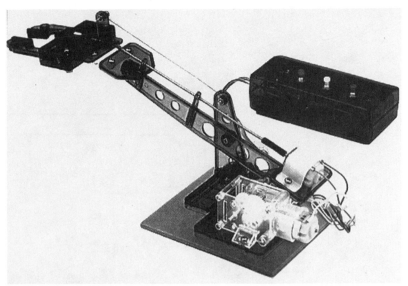

■ **15.5** *Stock OWl robotic arm*

This program turns the motor on clockwise (CW) for a second or so, turns it off, then turns it on in a counterclockwise (CCW) direction.

Timing a timing loop

There are many different kinds of IBM-compatible computers on the market operating at various speeds. The speed at which BASIC language executes determines how long a particular timing loop lasts, and the timing loop in the main program determines how long a motor stays on.

You want to find a timing loop number that turns on the motor for one second. With this determined, you can use multiples and fractions of this timing loop number to control the robotic arm to perform specific functions for specific periods of time.

For instance, suppose one second on your computer is equivalent to the number 10,000 in the timing loop. Turning clockwise one second would turn the motor on for a count of 10,000. An open gripper for $\frac{1}{2}$ a second would activate the gripper motor for a count of 5,000, and so on.

The second test program will provide a rough idea of this number. You can then tweak this number in the main program to achieve better results.

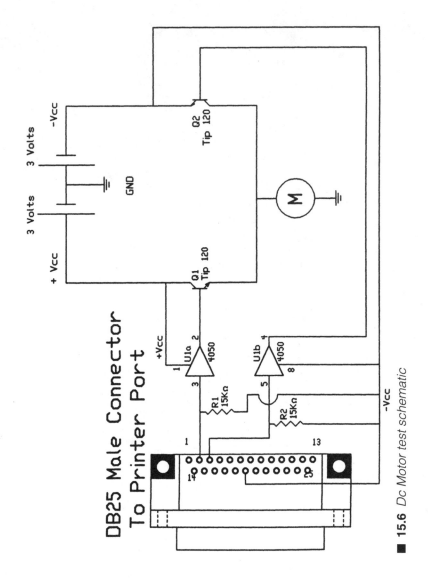

DB25 Male Connector
To Printer Port

+Vcc

U1a   2
4050
3

R1
15KΩ

U1b   4
4050
8
5

R2
15KΩ

-Vcc

1

14

13

25

+Vcc

Q1
Tip 120

GND

-Vcc

3 Volts

3 Volts

+ Vcc

Q2
Tip 120

M

■ **15.6** *Dc Motor test schematic*

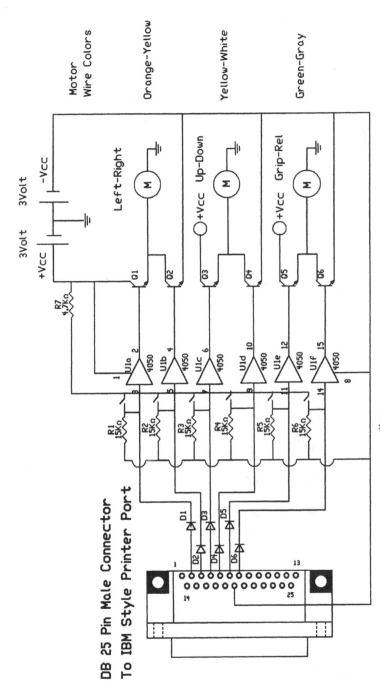

15.7 *Schematic for robotic arm interface*

Robotic arm

IBM interface

The complete robotic arm interface schematic is shown in Fig. 15.7. The PCB pattern for the circuit is illustrated in Figs. 15.8 and 15.9, and parts placement on the board is shown in Fig. 15.10. Figure 15.11 shows the top side of the finished interface.

Drill tape data			
Tool Code	Hole Size	Sym bol	Hole Count

Board type: FR4
Board thickness: .062"
Copper weight: 2oz

All dimensions are in inches.

Board should have green solder mask over bare copper (SMOBC) with white silk screening on top.

View is from the top side of the board.

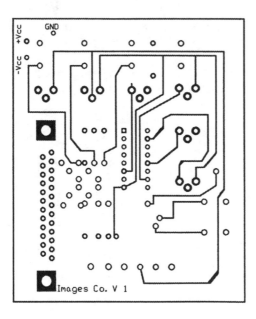

Top Side

Co:	?		
Title:	?		
Board:	?	Revision:	A
Drawn:	?	Size:	A
Date:	?	Sheet 1 of 1	

■ **15.8** *Top foil for PCB*

Drill tape data			
Tool Code	Hole Size	Sym bol	Hole Count

Board type: FR4
Board thickness: .062"
Copper weight: 2oz

All dimensions are in inches.

Board should have green solder mask over bare copper (SMOBC) with white silk screening on top.

View is from the top side of the board.

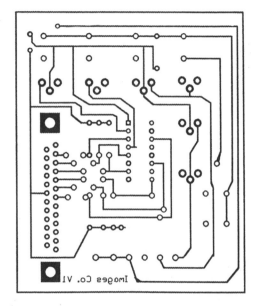

Imoges Co. V1

Bottom Side

Co:	?		
Title:	?		
Board:	?	Revision:	A
Drawn:	?	Size:	A
Date:	?	Sheet	1 of 1

■ **15.9** *Bottom foil for PCB*

Rather than solder the wires from the robotic arm directly to the interface board, we have used PC board wire terminals. This will allow you to change the length of the wires or remove the robotic arm for other projects with little hassle.

Wire the robotic arm motors to the circuit using the same sequence of wires as shown. Starting from the left, the wire colors are orange,

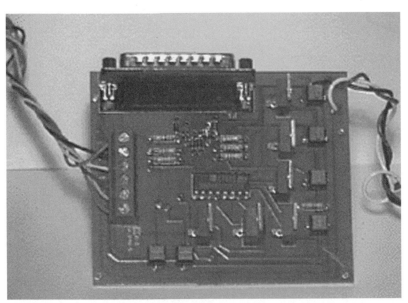

Push-Button Controls

GRIP   DOWN

RELEASE   UP

+Vcc GND

−Vcc

SW1   SW2   SW3   R7   SW4

Q1   Q2   Q3   Q4

Front View

Top View

Tip 120

Diode Close Up

D1 R1 R3
D2 R2
D3
D4
D5
D6

R4 R6 R5

4050

Q5

Q6

SW5   RIGHT

SW6   LEFT

Gray

Brown

White

Orange

Yellow

Green

■ **15.10** *Parts placement for interface*

■ **15.11** *Top view of finished interface*

yellow, green, white, brown, and gray. The last three wires (white, brown, and gray) may be put into any sequence since they all are connected to the circuit ground.

Connecting the interface to the computer

The IBM interface has a DB25 connector that can be directly plugged into the printer port, but it is much easier to connect the interface to the computer using a DB25 (M-F) cable. In most cases, the same port is also used for the printer. To alleviate switching cables back and forth whenever you want to use the robotic arm, purchase an A/B data switch (DB-25) box. Connect the robotic arm interface printer to the A side and the printer to the B side. Now you can use the switch to connect the computer to either the interface or printer (see Fig. 15.12).

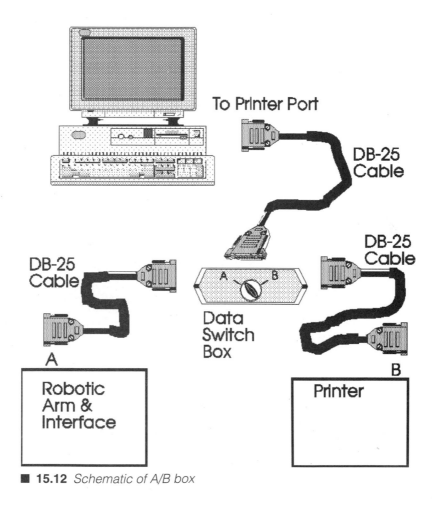

To Printer Port

DB-25 Cable

DB-25 Cable

DB-25 Cable

Data Switch Box

A

Robotic Arm & Interface

B

Printer

■ **15.12** *Schematic of A/B box*

Manual control

The robotic arm can be controlled manually using the interface, which has six push buttons. Each one controls a function. The manual controls may be operated at any time, and it doesn't make a difference if the interface is connected to the computer or not.

Joystick program

The joystick program allows you to control the robotic arm in real time through a joystick. The robotic arm will follow the movement called out by the joystick. Pushing the joystick left will cause the arm to rotate left, pushing the joystick right will cause the arm to rotate right, pushing the joystick forward will raise the arm, and pushing back will lower the arm. To activate the gripper, press the fire button. The gripper will alternate between gripping and releasing each time it's activated. The calibration section of the program will allow the arm to be used with a large variety of joysticks.

Keyboard program

The keyboard program allows one to control the robotic arm in real time using the keyboard. The following keys perform the following functions:

Key		Function
"U"	=	Up
"D"	=	Down
"L"	=	Left
"R"	=	Right
"G"	=	Grip
"H"	=	Release
"S"	=	Stop
"Q"	=	Quit

Programming the robotic arm

The main robotic arm program illustrates the programming of each robotic arm function. Each command is associated with a number that is ported to the printer port. For instance, to turn the arm left, the number 1 is sent to the printer port. This brings U1a high, which turns on transistor Q1 and activates the dc motor. Each command works in a similar manner.

If you read through the program, you will notice that only single function commands are given. It is possible to program two functions at a time, like move up (dr = 8) and turn left (dr = 1) by adding their dr numbers together and outputting it to the port (OUT a, 9). Dual function commands will increase the load on your power supply. If the power supply can't handle the load needed, the motor speed will slow down, not moving the full distance dictated by the timing loop. This becomes evident when programming repetitive operations used in animatronics. Each motion error is added to the last until the sequence fails. To prevent loading down the power supply, you may want to consider removing the battery power supply and replacing it with a Vdc wall transformer of adequate power.

Animatronics

You can program a repetitive sequence of movements for the robotic arm. This is a quick introduction in animatronics. The Main program already contains the code to perform repetitive functions. The REM statements from lines 130–145 are removed. The number assigned to variable rp determines how many times the robotic arm will perform the sequence of movements. In the program it's only two times but could very easily be 1,000 times or 10,000 times or even unlimited.

To program for an unlimited amount of repetitions, put the REM statements back in line 140. Now the program will continue the sequence indefinitely. To stop the program, hit the CTRL and Break keys.

If a sequence is run for any length of time, you may notice the position of the arm drifting from its original position. This will happen quickly if the battery power supply to the robotic arm is or becomes inadequate. The most effective way to deal with this is the use of positioning controls or limit switches (see below).

Going further

The first thing to consider is that the IBM controller circuit may be used with most hobby dc motors. Therefore, the controller may be used to control a variety of dc motors which may be employed in animatronic creations. A quick look at the printer port informs us that there are two unused pins. This is enough space for another dc motor controller.

With regards to the robotic arm itself, to advance this project, it would be helpful if the computer could determine the position of the arm. One way to accomplish this is with a potentiometer and an analog-to-digital (A/D) converter. A small, linear pot may be placed on the gripper section to determine closure, and a similar unit may be placed on the arm to determine its up and down position. A rotary pot may be used to determine sweep position as long as the arm is restricted to the travel of the rotary pot.

A simpler procedure may be used to determine absolute positioning. This would be a series of limit switches (momentary contact lever switches) that close when the arm reached its limit of travel in that direction. To reset the arm, have it proceed in the direction of travel until switch closure. The computer would then know the absolute position of the arm.

```
10 REM Use Timer to determine 1 second loop
20 ON TIMER(1) GOSUB 100
25 TIMER ON
30 FOR x = 1 TO 30000
40 NEXT x
100 BEEP
110 PRINT x
5 REM dc Motor Control Test
10 REM John Iovine 10-10-96
15 REM First find the Printer Port Address
20 DEF SEG = 0
25 a = (PEEK(1032) + 256 * PEEK(1033))
30 OUT a, 1: REM turn clockwise for awhile
35 GOSUB 200
40 OUT a, 0: REM stop for awhile
45 GOSUB 200
50 OUT a, 2: REM turn counterclockwise for awhile
55 GOSUB 200
60 OUT a, 0: REM STOP everything
65 END
200 FOR t = 1 TO 25000: NEXT t: RETURN
5 REM Programmed Robotic Arm Controller
10 REM John Iovine 10-10-96
15 REM First find the Printer Port Address
20 DEF SEG = 0
25 a = (PEEK(1032) + 256 * PEEK(1033))
30 DIM dr$(100), tr(100): REM direction and time35 REM commands
left, right, up, down, grip, release
40 REM grip = 16:release = 32:up = 8:down = 4:right = 2:left=
1:stop = 0
45 REM To Program Robotic Arm enter direction and time in data
46 REM statements
50 REM 16,000 = 1 second
55 FOR x = 1 TO 9: REM x equals number of data statements
60 READ dr$(x), tr(x)
65 c = tr(x): REM Set timing loop counter
75 IF dr$(x) = "right" THEN dr = 2
80 IF dr$(x) = "left" THEN dr = 1
85 IF dr$(x) = "up" THEN dr = 8
90 IF dr$(x) = "down" THEN dr = 4
95 IF dr$(x) = "grip" THEN dr = 16
100 IF dr$(x) = "release" THEN dr = 32
105 FOR i = 1 TO c
```

```
110 OUT a, dr: NEXT i
115 NEXT: REM next data statement
120 OUT a, 0
125 REM for repetitive action add following lines by removing
REM statements
130 REM rp = 2: REM RP = number of repetitive sequences
135 REM RESTORE: REM Reset DATA pointer
140 REM rep = rep + 1: IF rep >= rp THEN 150
145 REM GOTO 55
150 rep = 0: END
236 DATA right,16000 : REM turn right for 1 second
237 DATA left, 16000 : REM turn left for 1 second
238 DATA up, 8000: REM up for 1/2 second
239 DATA down,16000  : REM down for 1 second
240 REMCONDENSE DATA STATEMENTS
242 DATA up, 16000, down, 16000, up, 8000, grip, 8000, release,
8000
```

☐ OWI Robotic Arm Kit—$49.95

☐ IBM Printer Port Interface Kit—$8.95 (Kit includes 3.5"
diskette with programs)

☐ DB-25 Data switch box—$9.95

☐ DB-25 (M-F) Cable—$4.00

Parts list for arm interface

☐ Q1-Q6 Tip 120 NPN transistors

☐ R1-R6 15K-ohm $\frac{1}{4}$-watt resistors

☐ D1-D5 1N914 diode or equivalent

☐ 4050 hex noninverting buffer

☐ DB-25 right-angle connector

☐ PC board

☐ PC screw terminals (6)

☐ 4 C-cell battery pack

Available from:

Images Company
POB 140742
Staten Island, NY 10314
(718) 698-8305

Visa, Mastercard, and American Express accepted.

For shipping and handling, add $ 7.50. New York state residents
should add appropriate sales tax.

```
5 REM Keyboard Robotic Arm Controller
10 REM John Iovine 10-10-96
15 REM First find the Printer Port Address
```

```
20 DEF SEG = 0
25 a = (PEEK(1032) + 256 * PEEK(1033))
35 REM commands left, right, up, down, grip, release
40 REM grip = 1:release = 2:up = 8:down = 4:right = 16:left =
32:stop = 0
45 REM To Program Robotic Arm enter direction by pressing keep
50 CLS
55 PRINT : PRINT : PRINT " Press the following keys to activate
robot arm"
60 PRINT " U = Up  D = Down"
65 PRINT " G = Grip   H = Release"
70 PRINT " L = Left   R = Right"
71 PRINT " S = Stop   Q = Quit"
75 a$ = INKEY$
80 IF a$ = "u" THEN dr = 8
85 IF a$ = "d" THEN dr = 4
86 IF a$ = "g" THEN dr = 1
87 IF a$ = "h" THEN dr = 2
88 IF a$ = "l" THEN dr = 32
89 IF a$ = "r" THEN dr = 16
90 IF a$ = "s" THEN dr = 0
91 IF a$ = "q" THEN GOTO 100
93 OUT a, dr
95 GOTO 75
100 OUT a, 0
101 END
REM Joystick Robotic Arm Controller
REM John Iovine 10-10-96
REM First find the Printer Port Address
DEF SEG = 0
a = (PEEK(1032) + 256 * PEEK(1033))
REM commands left, right, up, down, grip, release
REM grip = 1:release = 2:up = 8:down = 4:right = 16:left = 32
REM stop = 0
SCREEN 0: CLS
GOSUB calibrate
start:
a$ = INKEY$
IF a$ = "s" THEN CLS : OUT a, 0: END
IF a$ = "q" THEN CLS : OUT a, 0: END
LOCATE 4, 15: PRINT " Use Joy Stick to control Robotic Arm"
LOCATE 6, 15: PRINT " Use fire button to activate gripper."
LOCATE 10, 15: PRINT "Press Q key to quit program"
REM ¥ coordinate (mean val = 72) (min val = 6) (max val = 173)
r = 15: dr = dr AND 15: LOCATE 16, 15: PRINT ""
IF STICK(0) < lj THEN r = 32: dr = dr AND 239: dr = dr OR r:
LOCATE 16, 15: PRINT "LEFT "IF STICK(0) > rj THEN r = 16: dr =
dr AND 223: dr = dr OR r:
LOCATE 16, 15: PRINT "RIGHT"
REM y coordinate (mean val = 76) (min val = 6)(max val = 171)
r = 51: dr = dr AND 51: LOCATE 14, 15: PRINT "  "
IF STICK(1) < fj THEN r = 8: dr = dr AND 251: dr = dr OR r:
LOCATE 14, 15: PRINT "UP  "
IF STICK(1) > bj THEN r = 4: dr = dr AND 247: dr = dr OR r:
LOCATE 14, 15: PRINT "DOWN "
putout:
OUT a, dr
IF STRIG(0) < 0 THEN GOTO lp1
LOCATE 12, 15: PRINT "  "
GOTO start
lp1:
IF flag = 1 THEN GOTO lp2
t1:
t = TIMER
```

236

```
s1:
flag = 1
r = 1: dr = dr AND 253: dr = dr OR 1: OUT a, dr
LOCATE 12, 15
PRINT "GRIPPER IS CLOSING"
 IF STRIG(0) < 0 THEN GOTO t1
IF (t + .15) > TIMER THEN GOTO s1
r = 62: dr = dr AND r: OUT a, dr
GOTO start
lp2:
t2:
t = TIMER
s2:
flag = 0
r = 2: dr = dr AND 254: dr = dr OR 2: OUT a, dr
LOCATE 12, 15
PRINT "GRIPPER IS OPENING"
IF STRIG(0) < 0 THEN GOTO t2
IF (t + .15) > TIMER THEN GOTO s2
r = 61: dr = dr AND r: OUT a, dr
GOTO start
calibrate:
PRINT : PRINT : PRINT "Do You Want To Calibrate Your Joystick
(Y/N)?"
q1:
a$ = INKEY$
IF a$ = "n" THEN GOTO ncal
IF a$ = "" THEN GOTO q1
IF a$ = "y" THEN GOTO q2
GOTO q1
q2:
CLS : LOCATE 12, 15
PRINT "Move joystick to left and hit fire button."
q3:
IF STRIG(0) < 0 THEN lj = STICK(0) + 10: GOTO q4
GOTO q3
q4:CLS : LOCATE 12, 15: PRINT "Release Joystick."
bt = TIMER + 3
WHILE bt > TIMER: WEND
cat = STRIG(0)
LOCATE 12, 15
PRINT "Move joystick to right and hit fire button."
q5:
IF STRIG(0) < 0 THEN rj = STICK(0) - 10: GOTO q6
GOTO q5
q6:
CLS : LOCATE 12, 15: PRINT "Release Joystick."
bt = TIMER + 3
WHILE bt > TIMER: WEND
cat = STRIG(0)
LOCATE 12, 15
PRINT "Move joystick forward and hit fire button."
q7:
IF STRIG(0) < 0 THEN fj = STICK(1) + 10: GOTO q8
GOTO q7
q8:
CLS : LOCATE 12, 15: PRINT "Release Joystick."
bt = TIMER + 3
WHILE bt > TIMER: WEND
cat = STRIG(0)
LOCATE 12, 15
PRINT "Move joystick back and hit fire button."
q9:
IF STRIG(0) < 0 THEN bj = STICK(1) - 10: GOTO q10
```

```
GOTO q9
q10:
CLS : LOCATE 12, 15: PRINT "Release Joystick."
bt = TIMER + 3
WHILE bt > TIMER: WEND
cat = STRIG(0)
LOCATE 12, 15
PRINT "Joystick is calibrated, press any key to continue!"
q11:
cat = STRIG(0)
a$ = INKEY$
IF a$ = "" THEN GOTO q11
CLS
RETURN
ncal:
lj = 25: rj = 80: fj = 25: bj = 80
RETURN
```

Android hand

16

THE LAST CHAPTER OUTLINED THE CONSTRUCTION OF A computer-controlled robotic arm. This chapter advances to construct a humanlike android hand. The actuator we will use to move the fingers in the android hand is the air muscle first introduced in Chapter 3.

The air muscle is a pneumatic device that produces linear motion with the application of pressurized air. Much like a human muscle, it contracts when activated. You may think, Well, pneumatic cylinders have been around for quite a while and do the same thing, but the air muscle is a boon to hobbyists and robotists because it is lower in cost, extremely lightweight, flexible, and easier (safer) to use.

The air muscle boasts a power-to-weight ratio of 400:1. Since most of its components are plastic and rubber, the air muscle can work when its wet or underwater. The flexible nature of the product allows it to function when bent around curved surfaces. These features make the air muscle the experimenter's choice over standard pneumatic cylinders.

Being pneumatic, the air muscle device operates off of air pressure, which is not as available as electric current. When I first encountered the air muscle, I thought that an air system would be too much of a hassle to build, regardless of how innovative the product. I was wrong; a simple (manual) air system can be put together for about $20.00.

Efficiency is lost when using electric power to compress air. However, the air muscle consumes little air volume per activation, and the compressed air can be stored. The air muscle's response and cycle time is fast. A small 6-inch, 10-gram air muscle can lift 6.5 lbs.

Before we build the android hand, we will first build a few manually operated air muscle demo devices. The demo devices allow you to become familiar with the operation and function of the air muscle before attempting a more complex project.

Manual control of an air muscle is fine for one or two air muscle projects. However, when five or six air muscles need to be operated in sequence or unison, manual control is not practical. For this we employ computer control. One may use a BASIC Stamp or IBM PC or compatible. The interface to either computer is the same. In this chapter we will use the PC. To control the air muscle via a computer (IBM or compatible printer port) through the PC's parallel port adds approximately $25.00 per air muscle to the cost.

Advantages of the air muscle

☐ Lightweight—6" air muscle with 18" of $\frac{5}{32}$" diameter air tubing weighs approximately 10 grams.

☐ Contraction—6" air muscle contracts approximately 1" (about 25% of its length, ends not inclusive).

☐ Powerful—Has lift approximately 6.5 lbs at 42 psi. Power-to-weight ratio can reach 400:1.

☐ Pliable—Soft, pliable construction. Can be bent around curved surfaces and still function properly.

Uses

The air muscle lends itself to robotics and automation. In some applications, it can replace servos and dc motors. Its unique properties (lightweight, strong, and pliable) can be capitalized in many applications and used to improve existing pneumatic designs. In a nutshell, the air muscle may be used in many applications that require linear or contractive motion. In many cases, pneumatic cylinders can be replaced.

How the air muscle works

The air muscle has a long tube constructed out of black plastic mesh. Inside of it is a soft rubber tube, and metal clips are fastened on each end. The plastic mesh is formed into loops on each end, tucked into and secured by the metal clips. The loops are used for fastening the air muscle to devices. When the air muscle is pressurized, the soft inner tube expands. The inner tube pushes against the black plastic mesh and causes it to expand also. When the plastic mesh expands, it shortens in length in proportion to the expansion of its diameter. This causes the air muscle to contract.

In order to operate properly, it is essential that the air muscle be stretched or elongated when it's deactivated or in a resting state.

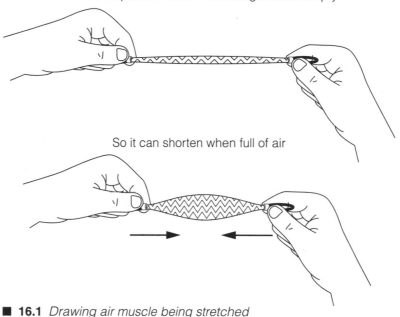

You need to pull the muscle out straight when empty

So it can shorten when full of air

■ **16.1** *Drawing air muscle being stretched*

If not, there will not be any movement or contraction when it is activated (see Fig. 16.1). When air pressure is applied, the muscle contracts.

Components of the air muscle system

Figure 16.2 illustrates the components needed to use the air muscle. Item 1 is the air muscle itself (of course). Item 2 is the three-way air valve, which allows one to manually operate the air muscle (see Fig. 16.3).

Item 3 is a bottle-top adapter with a pressure release valve (set around (60 psi). The bottle-top adapter allows one to use a standard plastic PET soda bottle for air storage. The pressure release valve automatically releases excess air when preset pressure is exceeded.

Item 4, PET soda bottle, is used for air storage. A plastic soda bottle can hold 50 psi easily. I have static-tested plastic PET bottles to 100 psi. CAUTION: NEVER USE ANY TYPE OF GLASS BOTTLE FOR AIR STORAGE. A slight fracture in a glass bottle or dropping it accidentally may cause the bottle to explode, sending tiny glass fragments all over. Plastic PET bottles elongate when overpressurized.

1) One standard 6" air muscle

2) Three-way valve
for controlling the
air flow

3) Bottle cap adapter
for attaching an air storage

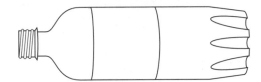

4) PET soda drink bottle
to be used as
the air storage

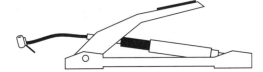

5) Foot pump adapter for
connecting a foot pump
to the $5/32$" air line

6) Foot pump

7) Nylon cable ties,
to attach the air
muscle to your
device

■ **16.2** *Items needed to experiment with the air muscle*

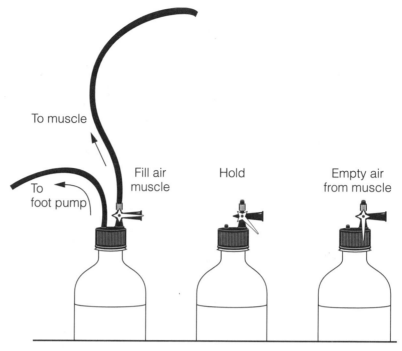

To muscle

To foot pump

Fill air muscle

Hold

Empty air from muscle

■ **16.3** *Three-way air valve, air muscle operation*

Item 5 is a foot-pump adapter, and item 6 is a foot pump. A simple foot pump with an air pressure gauge can charge air storage up to 100 psi. Because of the low volume of the PET bottles, air storage is brought to 50 psi with three or four strokes of the air pump. The air muscle uses so little air that a small PET bottle holds enough air for 4–5 complete cycles of the air muscle. Item 7, nylon cable ties, are used to quickly connect the air muscle to a mechanical device.

Figure 16.4 gives a general overview of how the parts are put together. In some cases you may want to epoxy glue some components together to prevent them from popping off. For instance, if you will just be using the three-way air valve on one bottle adapter for air muscle experiments, you may want to glue the three-way valve to the adaptor permanently.

Attaching the air muscle to mechanical devices

The air muscle is made of a soft inner tube encased in a strong plastic mesh. The assembly is held together by metal clips on each end, and the plastic mesh is looped at each end, making a hole. The plastic mesh hole is very strong mechanically and can be used

This is how they all fit together…

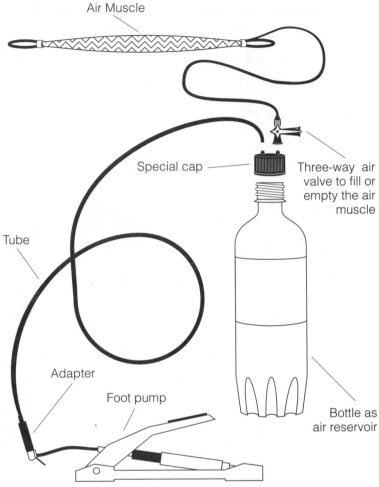

■ **16.4** *General overview, how parts fit together*

to attach the air muscle to any device. Figure 16.5 shows a machine screw inserted through the mesh hole.

Using the air pump adaptor

When you receive your foot pump, it will have a standard air nozzle as shown in Fig. 16.6. We need to replace the standard nozzle with the air-pump adaptor. Lift the locking lever as shown in Fig. 16.7. Remove the standard nozzle (Fig. 16.8), and insert the air-pump adapter (Fig. 16.9). Close the locking lever by pressing it back down.

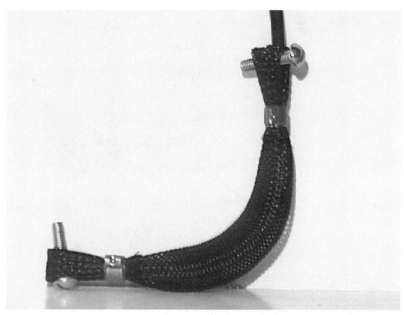

■ **16.5** *Screw going through one end loop of air muscle*

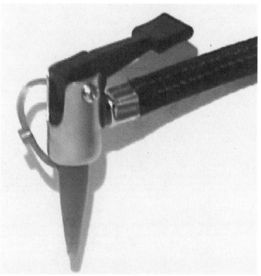

■ **16.6** *Foot-pump nozzle*

■ **16.7** *Lift locking lever (foot-pump nozzle)*

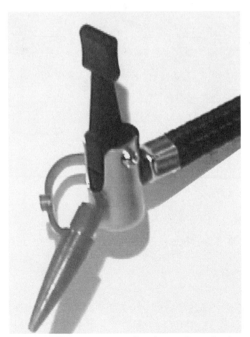

■ **16.8** *Remove standard nozzle adapter*

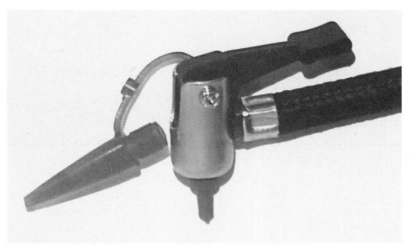

■ **16.9** *Insert air foot-pump adapter*

Have a Coke or Pepsi

You need to acquire a plastic PET bottle. The easiest way to do so is to buy a soda. Make sure the soda bottle is plastic, and don't purchase a soda larger than a 1-liter bottle. A half-liter bottle is ideal. I've tried the bottle-cap adaptor on all sizes of PET soda bottles up to 2 liters, and it fits all of them.

Empty the PET bottle of soda and clean it out with tap water. The bottle should be completely dry before using. It's interesting to note that if a full bottle of soda is dropped, the resulting pressure caused by the release of the carbonated soda greatly exceeds the 50-psi limit we impose on the bottle. Soda companies designed the PET bottle to withstand a rapid increase in bottle pressure that would come from dropping the bottle. This is something I never realized before working with the air muscle. Remember, no glass bottles should be used in the air muscle pneumatic system.

Building the first demo device

The first mechanical device we will build is a simple instrument that can be used to measure the contraction of the air muscle (see Fig. 16.10). The base is 1" × 2" lumber 11" long. I used this material simply because I have it lying around. You can just as easily use metal or plastic. At each end I drilled a hole to accept an 8-32 machine screw $1\frac{3}{4}$" long. The machine screws are inserted and held in place using two 8-32 nuts, one nut on each side of the wood. The head of the screw and shaft protrude about $\frac{3}{4}$" above the wood.

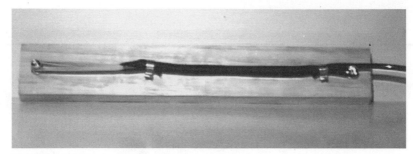

■ **16.10** *First mechanical device*

The top screw is threaded through the top opening of the air muscle before inserting the screw into the wood. A rubber band is looped through the bottom opening of the air muscle, which is looped around the bottom screw.

The rubber band stretches the air muscle when it is in its relaxed state.

Connect the balance of components as shown in Fig. 16.4. At times I've had difficultly connecting the $\frac{5}{32}$" diameter tubing to some of the components. Here are a few tips. First, if the tube refuses to go onto an adaptor, place the tubing under running hot water from the faucet. This softens the plastic, making it easier to fit onto the components. Another trick is to use some clear plastic tubing, which is snug enough to fit onto the adaptor nozzles properly (see Fig. 16.11). In addition, it is pliable enough to fit the $\frac{5}{32}$" tubing inside the tubing itself (see Fig. 16.12). The soft tubing acts like an adaptor and quick release for changing air muscle devices.

To operate the device, first pressurize the system using the foot pump. This only takes about four strokes to reach 50 psi. Your mileage may vary, depending on the size of the PET bottle you are using.

Open the three-way air valve to charge the air muscle. The muscle should immediately contract. You can measure the distance it moves in proportion to the psi gauge on the pump. You should be able to operate the muscle through four or five contractions and expansions before you need to refill the PET bottle. The air muscle doesn't use much air.

Notice that the air muscle stays in the contracted position until the three-way valve is turned to release the air from the muscle. It doesn't cost any energy to keep the air muscle contracted, in contrast to servo motors and solenoids that must be supplied electrical energy continuously to maintain their push or pull.

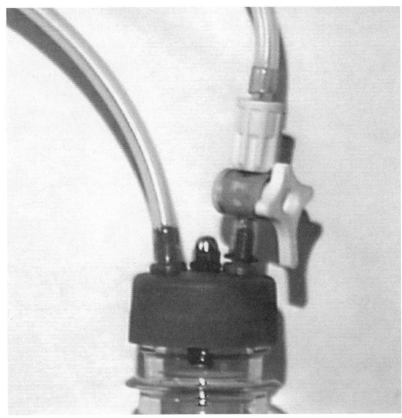

■ **16.11** *Using clear plastic tubing on standard adapters*

■ **16.12** *Using clear plastic tubing and ⁵⁄₃₂" tubing*

If the muscle doesn't appear to contract, then it probably wasn't stretched far enough. Remember, the muscle must be stretched in order for it to contract (operate).

Second mechanical device

The second device is a lever (see Figs. 16.13 and Fig. 16.14). The lever I made is constructed out of wood and plastic. Machine

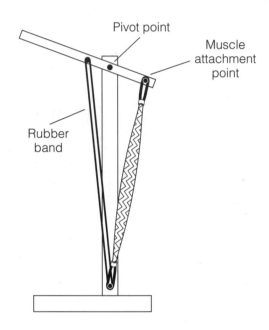

When the air muscle fills, it shortens, pulling the lever up.

When the air is let out, the muscle lengthens and the elastic bands pull the lever down.

■ **16.13** *Second mechanical device "lever"*

■ **16.14** *Second mechanical device "lever"*

screws secure the air muscle and rubber bands to the lever arm, and a wood screw through the plastic arm is the pivot. A second wood screw holds both the air muscle and rubber band. Operate this device using the three-way air valve as before. When activated, the lever moves up.

IBM interface

Computer control is easy. The computer operates an electrical three-way air valve. There are two inexpensive three-way electrically operated solenoid air valves available (see Fig. 16.15). These air valves operate at either 12 or 24 volts dc. The 12-volt dc air valve is the cylinder shape and is rated at 50 psi. The rectangle air valve operates at 24 Vdc and is rated at 100 psi.

To operate a single air valve, we only need one pin off the parallel (printer) port, along with a ground (see Fig. 16.16). The pin is a buffer with a 4050 noninverting hex buffer. The output of the hex buffer turns on or off a TIP 120 NPN Darlington transistor, and the transistor controls the current going to the air valve.

Basic program

The basic program is short and simple. After finding the printer port address, the subsequent lines control the valve of pin 2. By

■ **16.15** *Two electrically operated three-way air valves*

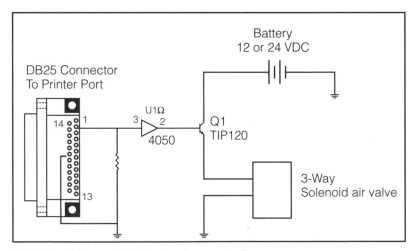

■ **16.16** *Schematic IBM air valve controller*

bringing the DB25 pin 2 high, the air valve is opened, allowing air pressure to the air muscle. Bringing pin 2 low closes the air valve to the muscle and vents the air from the air muscle.

More air

The air muscle, as previously discussed, uses compressed air from a PET plastic air storage bottle and foot pump. One can use compressed air from just about any source that's available. For instance, you can purchase small compressed air canisters used for air brushing. In fact, air-brushing supplies may provide you with a list of suitable tubing and fittings to experiment with.

There are a few small air compressors available on the market. The more expensive ones, made for air brush painting, have metal storage containers and air pressure regulators. At the other end of the market are the inexpensive 12-Vdc portable air compressors used for tire inflation. These compressors typically do not have an air pressure regulator or air storage. However, these items may be purchased to make an inexpensive pneumatic system. Do not use plastic PET bottles for air storage with any kind of automatic air compressor. The plastic PET bottles are only suitable for air storage with hand (or foot) operated air pumps.

Safety first

Since this is a new product, not too many people may be familiar with working with pneumatic systems. Therefore a few safety guidelines should be followed.

1. Always wear goggles when prototyping a new design.
2. Never connect a plastic PET soda bottle to an air compressor.
3. Never use a glass bottle for air storage.
4. Limit PET bottle size to one liter (or quart) or less.
5. Do not unscrew the bottle top; pull off an air fitting or valve when the system is still pressurized. Bleed the system of air first.

Android hand

The construction of a human-type hand begins with a trip to a toy store. The toy needed is called an Awesome Arm (TM) made by Zima Company in China (see Fig. 16.17). You will need to purchase two Awesome Arms to get enough fingers. The thumb on the toy is fixed and cannot be used.

The toy works by allowing the users to use their own fingers to actuate fingers on a robotic hand—something like an extension hand. We will scavenge the main component out of this toy to make an inexpensive android hand.

When you turn the arm over, there are five small screws that hold the hand together. Remove the screws, and the arm comes apart (see Fig. 16.18). Remove the finger section of the toy (see Fig. 16.19) and discard the rest of the components. The rectangular boxes on the finger pulls are where a person places his or her fingers to use the extension hand. We won't be needing them, so cut them off using wire cutters, leaving a long plastic stem.

We need to build a substructure to support all of the components. I began by tracing the outline of my right hand on paper. Then I shaded in a form that would become the support structure (see Fig. 16.20). The shaded area was cut from $\frac{1}{8}$"-thick aluminum plate.

The fingers must be secured to the end of the plate. To do so, first mark the position on the support aluminum. Next place a small aluminum plate $\frac{1}{2}$" wide × $\frac{1}{8}$" thick right behind the plastic back of the fingers (see Fig. 16.21). This forms a back plate for the finger base to rest against. Drill three holes through the two aluminum plates

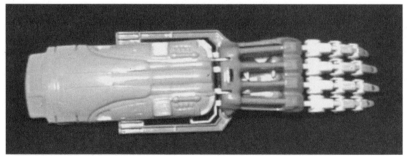

■ **16.17** *Awesome arm manufactured by Zima Co.*

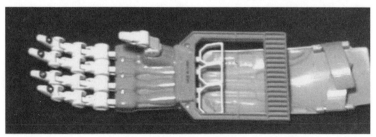

■ **16.18** *Opposite side of arm, where screws are removed*

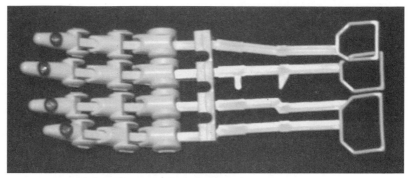

■ **16.19** *Finger pulls salvaged from arm*

Android Hand Outline

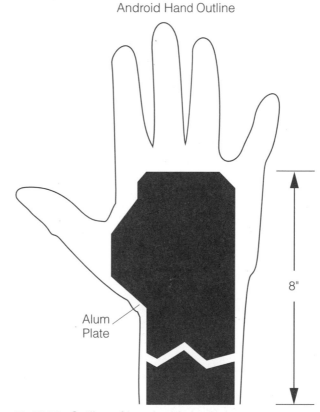

Alum
Plate

8"

■ **16.20** *Outline of hand and aluminum support*

and fasten the small plate in position using a few machine screws and nuts.

A top aluminum plate ($\frac{1}{8}"\times\frac{1}{2}"$) is secured to the top of the finger base. Drill four holes through the top plate and support plate as shown by the screw positions in Fig. 16.22. Four 1" -long machine

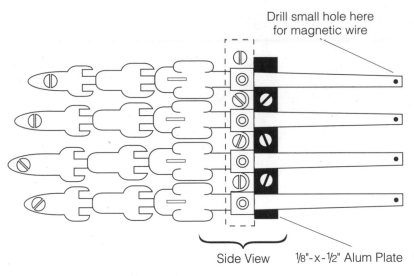

■ 16.21 *Placement of back plate.*

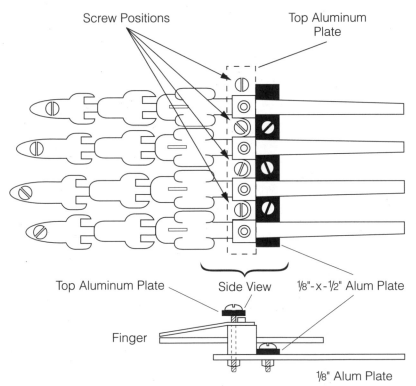

■ 16.22 *Placement of top plate.*

screws and nuts secure the top plate in position. These machine screws serve a dual purpose. First they secure the top plate that locks the fingers onto the support plate. Second, they will hold a rubber band to provide tension for the air muscle.

With the fingers secured to the support plate, we need to attach an air muscle to each finger. In order for the air muscle to provide a useful contraction, it must be stretched. A rubber band is looped through the air muscle. The first of the four 1" screws securing the top plate is removed. The looped ends of the rubber band are placed where the screw passes, then the screw is replaced, threading through the looped ends through the top plate holes and secured with a nut (see Figs. 16.23 and 16.24).

The air muscle is pulled back until it's fully extended, and the end of the air muscle is marked. A hole is drilled on the mark and the machine screw and nut are placed there. The end loop of the air muscle is placed over the machine screw holding the air muscle in an extended position (see Fig. 16.25).

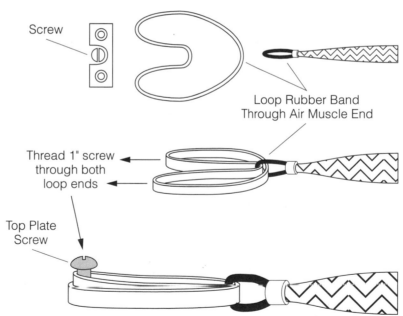

Screw

Loop Rubber Band
Through Air Muscle End

Thread 1" screw
through both
loop ends

Top Plate
Screw

■ **16.23** *Threading rubber band through one end of air muscle and attaching opposite end to top plate screw*

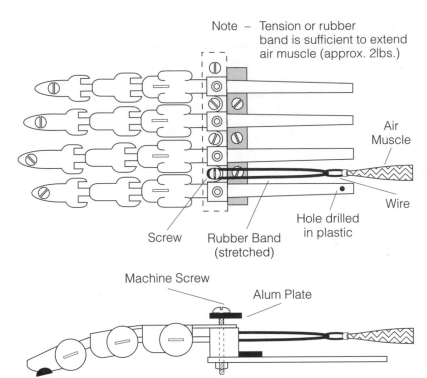

Note – Tension or rubber band is sufficient to extend air muscle (approx. 2lbs.)

Air Muscle

Wire

Hole drilled in plastic

Screw

Rubber Band (stretched)

Machine Screw

Alum Plate

■ **16.24** *Overview, attaching stretched air muscles to finger pull*

Now a small hole is drilled in the plastic section of the finger pull. The small hole should line up with the front loop of the air muscle, and the hole must just be large enough to pass a double strand of magnetic wire. You can substitute bare 22-gauge solid copper wire for magnetic wire.

A double strand of wire is passed through the plastic hole and front loop of the air muscle. The ends of the wire are twisted together, securing the components together. If there is excessive wire left from twisting, clip it off using wire cutters.

The top view should look something like Fig. 16.24. We can now see how the finger will contract. As the air muscle is pressurized, it contracts. The contraction pulls the plastic stem of the finger pull, which in turn contracts the finger. When pressure from the air muscle is released, the rubber band extends the air muscle back into its original extended position.

At this point it's a good idea to static test the finger. Connect the air supply to the muscle to ensure that it operates in the manner just described. The prototype required a pressure of 42 psi to fully contract the index finger.

When the finger operates properly, connect the air muscles to the remaining fingers in the same manner described. Figure 16.26 is a closeup of air muscles connected to all of the finger pulls.

Air Muscle Loop

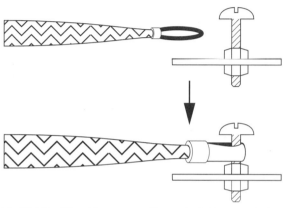

■ **16.25** *Attaching opposite end of air muscle to maching screw to extend air muscle*

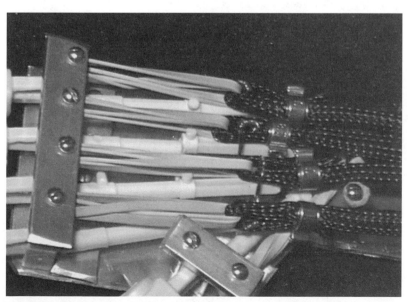

■ **16.26** *Close-up of air muscle, rubber band, and finger pull tied together in finished hand*

The thumb

The thumb is the most important finger on the hand. It makes grasping, holding, and using tools much easier. If you don't think so, try picking up a coin off a table or floor without using your thumb. Now try using a few tools, like pliers, wire cutters, a hammer, or drill.

To make the thumb, cut off the small finger assembly from the second hand unit purchased. Assemble this finger section lower and at a 45-degree angle to the other fingers (see Fig. 16.27).

The thumb in this prototype is articulated (moves) but is not opposable. To improve this design, make the thumb opposable. This will increase the effectiveness of the hand.

To make the thumb opposable, the thumb-containing portion of the hand must be cut off and reattached using a spring-loaded hinge (see Fig. 16.28). The spring-loaded hinge would be located on the rectangular box shown in Fig. 16.28. An air muscle connects to this section, and when activated pulls the thumb into the palm section of the hand. This makes the thumb opposable and articulated.

Going further

The robotic hand can be interfaced to an IBM-compatible computer using five electrical solenoid valves, similar to the single-valve design shown earlier. An outer covering like a rubber hand can be fitted over the robotic hand to make an android hand (see Fig. 16.29).

Some other applications found for the air muscle are interesting. Here are a few;

- ☐ Six-legged robotic walker
- ☐ Easy-open jar clamp for people with arthritis

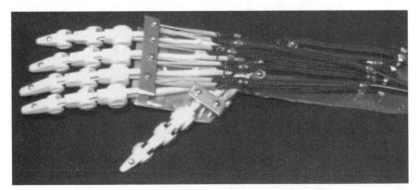

■ **16.27** *Finished robotic hand*

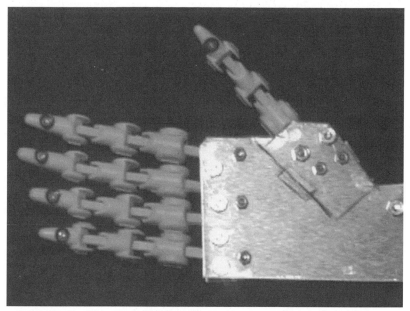

■ **16.28** *Plans to make thumb opposable as well as articulated*

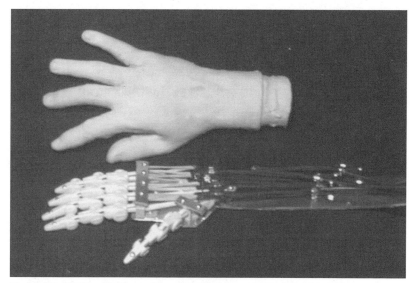

■ **16.29** *Fitting lifelike rubber hand over robotic hand to create an android hand.*

- [ ] Robotic hands
- [ ] Robotic arms

Basic program

```
5   REM Solenoid Air Valve Controller
10  REM John Iovine
15  REM Find Printer Port Address
20  DEF SEG = 0
25  a = (PEEK(1032) + 256 * PEEK(1033))
30  REM Next line activates the air muscle
35  OUT a,1
40  REM Next line deactivates the air muscle
45  OUT a,0
```

Air muscle parts list

- [ ] Air Muscle 6" length with $\frac{5}{32}$" tubing—$15.95
- [ ] PET bottle-top adaptor with pressure release valve—$4.00
- [ ] Three-way air valve—$4.00
- [ ] Air-pump adaptor—$2.00
- [ ] Foot air pump with 100-psi air-pressure gauge—$12.95
- [ ] $\frac{5}{32}$"-diameter air tubing—$.25 per ft
- [ ] $\frac{7}{32}$"-diameter clear air tubing (for making quick releases)—$.25 per in

IBM interface

- [ ] 12-Vdc three-way solenoid air valve, 50 psi max.—$20.00
- [ ] 24-Vdc three-way solenoid air valve, 100-psi max.—$20.00
- [ ] DB25 pin connector—$3.50
- [ ] 4050 Noninverting hex buffer—$1.00
- [ ] TIP 120 NPN Darlington transistor—$1.25

Parts available from:

Images Company
POB 140742
Staten Island, NY 10314
(718) 698-8305

Visa, Mastercard, and American Express are accepted.

Index

263

Illustrations are in **boldface**.

265

269

270